QUEEN BREEDING AND GENETICS
How to get better bees

written and illustrated by

EIGIL HOLM

Northern Bee Books

Queen Breeding and Bee Genetics

© Eigil Holm 2010

ISBN 978-1-904846-62-8

Translated from the second Danish edition 2009 by Eigil Holm

Lay-out and illustrations: Eigil Holm

43 photos and 65 drawings and schemes by Eigil Holm are not marked by name.
Other sources: Danish Beekeepers' Association (fig. 60), Hans Røy (fig. 61), Queen Breeders' Association (fig.78), Demerec and Kaufmann (fig. 85), Charlie Christensen (fig. 90), Brother Adam (fig. 112), Bo Vest Pedersen (fig. 114).

Photos: Danish Beekeepers' Association (2a,b), Swienty (fig. 17), Per Kryger (fig. 110).

Parts of this book must only be reproduced with the written permission of the editor.

Republished in the United Kingdom by
Northern Bee Books,
Scout Bottom Farm,
Mytholmroyd,
West Yorkshire HX7 5JS
Tel: 01422 882751
Fax: 01422 886157
www.GroovyCart.co.uk/beebooks

Other books of Eigil Holm, see www.eigilholm.dk

Content of Queen Breeding and Bee Genetics

PREFACE	5
1. PRODUCTION OF QUEENS	6
The basis of queen production	6
Methods of queen rearing	7
Grafting	8
The calendar	8
The cell builder	9
The breeder colony	10
Grafting tools	10
Grafting procedure	11
Transfer to the cell builder	12
Incubation	12
How to obtain young bees	14
Marking of queens	14
Transport cage and candy	16
Shipping	17
2. MATING NUCS	18
How to use the nuc	19
Continous management of a nuc	21
Clearing after the mating season	21
3. ALTERNATIVE METHODS	22
1. The queen production and breeding of Poul Erik Sørensen	22
2. Queen right cell builder	24
3. The starting box method	25
4. MATING STATIONS	26
Organization of the queen breeding	26
Drone producers	28
Placing of the mating nucs	29
One-comb nuc	30
How to remove drones	31
Transport	31
How to treat the mated queens	31
5. INSTRUMENTAL INSEMINATION	33
Types of insemination	33
Insemination in genetic research	33
On techniques of insemination	33
Drones and their sperm	34
6. HOW TO GET BETTER BEES	38
The ABC of breeding	38
1. The aims of breeding	38
2. Judgment	38
3. Selection	38
4. Crossing	39
6. Maintenance	39
7. AIMS OF BREEDING AND JUDGMENT	40
Marks used for judging bee colonies	41
Comments to the marks	42
Marks of the year	43
How to use the marks	43
Other judgments	43
Judges	44
Environment and judgment	44
8. QUEEN BREEDING AND GENETICS	45
The genetic consequences of the mating behaviour	46
Congregational areas	47
9. HOW TO MAKE A PEDIGREE	48
The danish breeders pedigree	49
10. BEE CYTOLOGY AND GENETICS	50
The cell	51
The genes	51
The anatomy of the gene	53
The genome and how to read it	53
Microsatellites	55
Maternel and parentel heritage	56
Mitochondria	57
11. MEIOSIS AND GENES	58
Dominant and recessive alleles	58
Mating and segregation	60
Mating in populations	62

Co-dominance	62
Mendel's firs law	62
12. MENDEL'S SECOND LAW	**64**
Segregation of two gene pairs	64
Why are bees different	66
Linkage and cross-over	66
Mendel´s second law	68
13. SEX ALLELES, ADDITIVE GENES, NATURE AND NURTURE	**69**
Effect of the number of sex alleles	70
Additive genes	71
Mutations	71
Inbreeding and heterosis	73
Nature and nurture	73
The relative importance of genes and environment	74
Heritability	74
14. BEE RACES AND EVOLUTION	**75**
The origin of races	76
Nature is changing	77
Races	78
The fate of the races	78
How to use the races	79
Morphological characters of races	80
Genetic examination of races	80
15. BREEDING	**82**
The Buckfast bee	82
Combination Breeding, an overview	82
Selection	83
Crossing strategy	83
Testing	84
Maintenance of stocks	85
Stocks for drone production	86
Closed populations	86
16. A GERMAN BREEDING SYSTEM	**87**
17. THE FUTURE	**89**
LITERATURE	90
INDEX	91

Preface

Breeders and producers of queens need a handbook which describes the background of their work. It can only be done if you know the techniques of rearing queens and judging the results. You must know the genetics of the bees, too, because it enables you to plan the maintenance and improvements of your stock. The aim of this book is to help the practical people who rear queens.

The chapter on instrumental insemination describes the general principle of that important work. There exist many different insemination apparatuses and you have to learn their use by studying the manuals. The laboratory work should be learned through a course.

This book was edited in 1995 in Danish and 1997 in German. After the genome of bees was described in 2006 and the methods of studying races and ecotypes by using microsatellites was developed a new edition was needed. It came out in Danish in 2009 and was translated and prepared for the English speaking market at the same time.

I thank dr. Per Kryger, leader of the bee research at the University of Aarhus and dr. Bo West Pedersen of the University of Copenhagen. Both of them have read the manuscript and advised me. Poul Erik Sørensen, the greatest producer of queens in Scandinavia and inventor of new breeding methods have gone through the chapters on practical breeding and permitted me to describe his methods. The board of the Danish Association of Queen Breeders have read all of the manuscript. I attended a course on instrumental insemination in Ohio by Susan Cobey. We had fine talks about queen rearing. I thank all of them heartily for their advice.

Gedved, January 2010
Eigil Holm

1. Production of queens

The queens live 3 or 4 years if all goes well. However, the beekeeper will often change his queens when they are one or two years old, or the bees will do it. Many beekeepers have experienced that 3 or 4 year old queens, and sometimes even 5 year old ones, perform just as well as the young ones. At which age should you change your queens? That is the decision of the beekeeper. You are free to choose if you have reserve queens in small families.

You can buy queens from local producers and imported ones from other countries, depending on legal restrictions. The beekeeper can produce his own queens, some methods are described below. The production of queens is not difficult but there are many steps in the work and you have to do the work exactly after the calendar. The beekeeper must follow the manual strictly, otherwise he will fail.

Fig. 1. Swarm cells are numerous and are often placed around the site of the brood.

We distinguish between production and breeding of queens. The aim of production is just to get more queens. The aim of breeding is to get better queens, they have a pedigree and you know from which queens the drones for mating come.

The basis of queen production

Queen production is based on what we know about the biology of bees, particularly the factors leading bees to treat the larvae so they become queens. It occurs in three cases: When bees prepare to swarm, when the queen dies, or if she is weakened.

The preparations for swarming happen when the family grows too big or has too little room. In colonies with a strong tendency to swarm it occurs on a definite time of the year, nearly independent of the conditions of life. The bees build swarm cells, often at the edge of combs with brood (fig. 1), and after their sealing the swarm leaves the colony. The swarm cells can be cut out and used to start new colonies. This method is not used in modern beekeeping because it makes the tendency to swarming greater which is exactly the opposite of what you desire.

Emergency cells are built when the colony feels itself queenless. Then the bees rebuild cells with young larvae of the right age. The colony might loose its queen if the beekeeper wounds her by accident, or she is damaged by bees clustering around her, or she gets a disease.

Emergency cells are often randomly distributed on combs. The number is irregular, often between 10 and 25 (fig. 3).

The bees have only a short time to make new queens after the death of the old one. The larvae must be less than three days old if they shall become good queens. If they are 3 days old they have developed so many worker's characters that they cannot be queens. Instead, they become intermediate between workers and queens.

Most methods of queen rearing have their ba-

Fig. 2. Supersedure cells are few but big. One or a small number are placed on or above the brood comb. (Danish Beekeepers Ass.).

sis in what we know about emergency cells. The beekeepers takes the queen away from a colony. A couple of hours later he gives it young larvae in cups on which the bees build queen cells. In most cases the bees provide them with royal jelly and they become queens of good quality if the colony has got food enough.

Supersedure. When the queen fails to produce enough queen pheromone the bees build a small number of queen cells, often 4 or 5, sometimes less (fig. 2). The old queen lays one egg in each of the cells. One of the queens from these cells replace the old queen. During some days or weeks the two queens can be found in the same colony.

The beekeeper uses the state of the colony which results in supersedure when he places queen larvae in a super above a queen excluder. Here the distance from the rest of the colony is so big that the bees in the super find supersedure necessary. Therefore they take care of the queen larvae so that they become queens.

Methods of queen rearing

The most common method is *grafting* which is described below. This method is more advantageous than other methods because the beekeeper is able to control all the steps from egg to mated queen.

Fig. 3. Emergency cells are placed at random depending on the place of the larvae which the bees choose.

1. PRODUCTION OF QUEENS 7

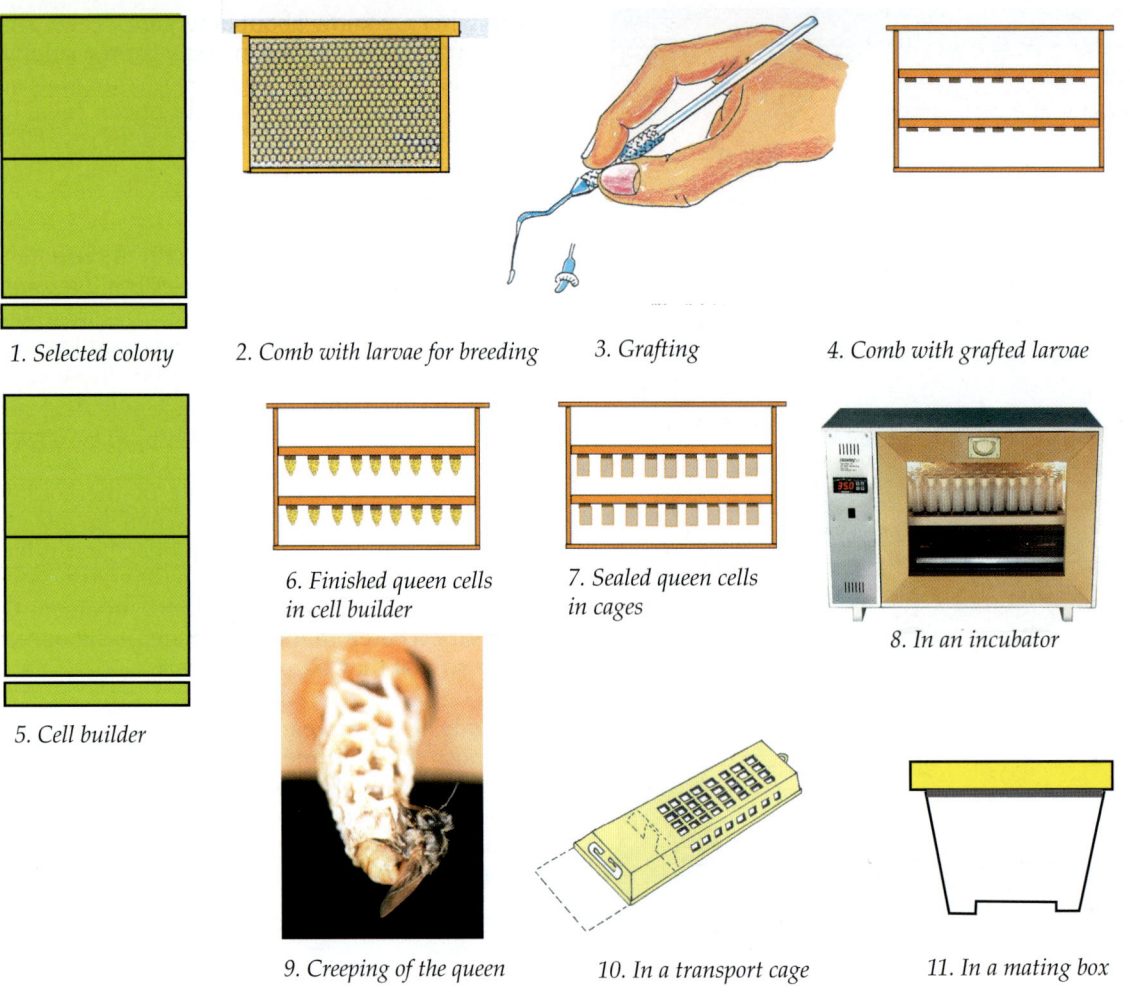

Fig. 4. Queen rearing, scheme.

Grafting
By grafting you take young larvae from a colony and places them in artificial cell cups. The larvae are taken from a selected colony. The grafted larvae in their artificial cell cups are placed in a queenless colony called the *cell builder*. Here they stay until they are sealed. After that they are transferred to an *incubator*.

Here they emerge in separate *cages*. From there they are transferred to a small colony in a *nucleus box (nuc)*. Here the queens remain until they have mated and begun the egg laying. Finally they are placed in a colony where they replace the old queen.

The many steps of the work is shown in fig 4.

There exist other methods for producing of queens. The grafted larvae are placed in a queenless *cell builder (starter)* which only starts the feeding of the larvae. Then they are transferred to the *cell finisher*, here they are nursed until the cells are sealed. After that they are put into an *incubator*. Instead, they can be placed in an incubator *colony* where they emerge in cages.

The calendar
The work is strictly controlled by the calendar.

In Denmark the queens are mated on the mating stations in June, July and until mid August. The cell builders must be ready to receive queen cells about the 20th of May and it takes 9 to 14 days to bring them so far that they can nurse the queen larvae. Grafting before that date can give bad results. (Scheme page 9).

It takes about 40 days from a drone egg is laid

CALCULATION OF DATES FOR BREEDING		
day no.	day no. since creeping	work
0	− 12	grafting of young larvae at 10 a.m.
5	− 7	into incubator
12	0	breeding at 10 p.m.
17	+ 4	to mating station
31	+ 18	back from mating st.

until the drone is ready for mating. Drones which shall mate in the beginning of June come from eggs which are laid from 15th to 20th of April.

It takes 16 days from the laying of the egg until the young queen leaves her cell. 6 or 7 days later she is ready for mating. The queen begins to lay eggs some days after mating. That is the control of a successful mating.

The planning has to be planned date by date. The creeping of the queen must be planned within 3 hours. (Scheme above)

The works are done in this order:

1. The colonies which deliver the queens are controlled. The final selection of them occurs during the first half of May.

2. The cell builder is made ready.

3. The grafting is done the 20th of May.

4. The queens emerge the 1st of June.

5. After that they are placed in a nuc, or a one comb cassette if it is going to be sent to a mating station. The first mating occurs the 7th of June or later depending on the weather.

6. End of June: Control of egg laying. Now every egg laying queen can be transferred to the colony where she shall live.

The date for later graftings is decided upon by the planning of the mating station.

The cell builder

The cell builder is described according to the method of Brother Adam, modified by Anders Glob. This method is good for beekeepers who want to produce a limited number of queens. Methods for large scale production is described in chapter 3.

The cell builder must be a very strong colony which is able to nurse the queen larvae. It is done by uniting two colonies, giving them little room, and feeding them well. Easier methods exist but they are not so sure in all types of weather as this. It can rear 100 to 110 queens in two teams in 10 days.

Two weeks before grafting two strong colonies are united. The queen is taken away from colony no. 2 and the magazine with that colony is placed on top of colony 1 with a queen excluder in between. Colony 1 keeps its queen. Every evening the united colony is fed with from 0,5 to 0,75 litres of sugar water (40 – 45% sugar). Fig. 5 A.

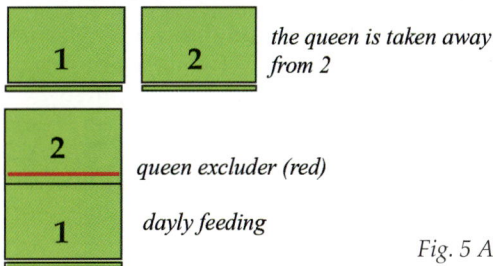

Fig. 5 A

5 or 6 days before grafting the super is inspected for emergency cells which are removed. It is necessary to shake all of the bees off the combs to be sure that there are none. The next 4 or 5 days the colony is not inspected but the feeding continues (fig. 5B).

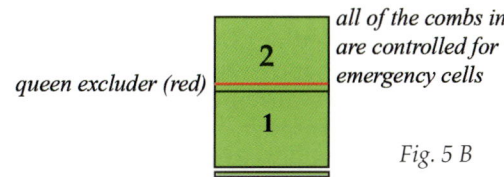

Fig. 5 B

Few hours before grafting the room for the bees is reduced such that all of the bees are in box 1. It is done thus: Box 2 is placed beside the hive. The queen is found in box 1 and is removed to a box in another place of the bee yard together with the comb on which she is found, a pollen comb and a piece of sugar candy. The size of the entrance is reduced to prevent robbery. (fig. 5 C).

1. PRODUCTION OF QUEENS

Fig. 5 C

After that box 1 is removed. A queen excluder is placed on the bottom to prevent other queens from entering (fig. 5 D). Now box 2 is placed on the queen excluder and 6 combs with brood (yellow) and 2 heavy pollen combs (blue) from box 2 are placed as shown on the figure. The bees from box 1 are shaken out in front of box 2. They march in.

Brood combs from box 1 are distributed among the colonies of the bee yard.

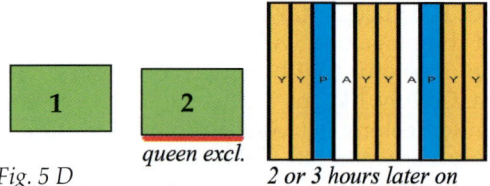

Fig. 5 D

Wait for a couple of hours and transfer two frames with grafted larvae in box 2 (white) as shown on fig. 5 D The cell builder is able to nurse 2x30 larvae. When they have been sealed after 5 days you can place new frames with 2x30 larvae. If the honey flow is too small you need to feed. The reason why the colony cannot produce more queens is the lack of young bees.

If you have done the work well about 80% of the larvae will become queens.

After the second team of larvae has been removed you transfer the old queen to box 2 in a cage. You could also let a queen cell remain. That queen will emerge inside the colony and become the new queen.

The breeder colony

You need to compare many colonies to find the best ones for breeding. The selection of colonies is described in chapter 7. It is based on big honey production, low swarming tendency, gentleness, resistance to diseases etc.

A condition for successful breeding is that you record the results of your inspections every time you examine the colonies.

The day of grafting you find a comb with many young larvae of the same age. From this you pick the larvae. It can be a time consuming work to find the larvae. Instead, you enclose the queen by queen excluders on both sides of a comb with empty cells 4 or 5 days before grafting. On both sides you need a comb with young larvae so that

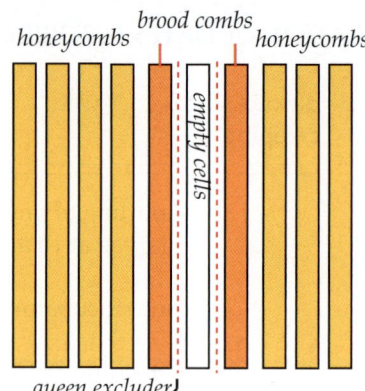

Fig. 6. Eggs of the same age are obtained by enclosing a queen on a comb with empty cells using queen excluders. A brood comb with larvae are placed on either side. The task is done in a super among honey combs.

the queen can be nursed and her comb can stay warm (fig. 6).

The queen lays her eggs in the comb. After 24 to 36 hours the queen is allowed to go to the other combs. The queen excluders remain so that you are sure there will not be laid more eggs in the comb. Then you have many larvae of the same age for grafting.

4 or 5 days after the queen was confined you can do the grafting. Now the larvae have used three days as eggs and lived 12 to 24 hours as larvae which is the right age for grafting.

If you have no breeder colony you can get a piece of comb with larvae from another queen breeder. The larvae tolerate transportation for up to 48 hours if the comb is enclosed in moist cloth within a plastic bag. They tolerate cooling better than heat (above 35^0).

Grafting tools

The grafting tools are:
grafting needle
cell cup
cup container
container holder
frame with bars for the container holders
royal jelly
lamp with cold light, for example diodes

The grafting needle is made by stainless steel (fig. 7). It has a flat part at its end for the larvae. You can use it as a ruler for the larva, which should not be more than 2 millimetres long. An egg is 1,5 millimetres long, it can always be found.

Cell cups are made of beeswax or plastic, the material is unimportant for the bees.

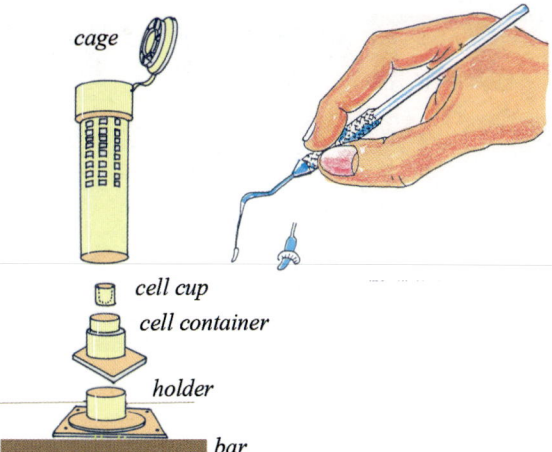

Fig. 7. Tools for grafting. To the right: The needle in the hand and the flat part of the needle with a larva of the right size. To the left: Exploded drawing of the cup and its fixation to the wooden bar. The cage is placed on the cup container 24 hours before the queen emerges.

Cup containers are made of plastic (fig. 7). The plastic cell cups fit in them, glue is unnecessary. The cell containers fit in a holder which is fixed to a wooden bar with nails. There are room for 8 to 11 holders on a bar. Two or three bars are placed in a frame. The vertical distance between bars is 90 mm. The bar should be 10x10 mm in cross section. Then the bees are less liable to build honey cells between the bars.

A cage for the new queen can be fixed to the cup container (fig. 7).

Grafting can be done outdoor in shadow. The sunrays kill the larvae, if they are exposed to it for too long time. But you might let the sun shine through the cell walls to get sufficient light. If you graft indoor you need lamps with cold light. Light from an electric bulb is warm and destroys easily the larvae.

After every grafting the larvae in the comb and the cups are covered by moist cloth to prevent drying-out. The larvae are very sensitive because they are small.

Grafting procedure

Before you begin to graft you take away a comb in the cell builder. Here you will place the grafted larvae. In a couple of hours the empty space is filled up with young bees ready to nurse the larvae.

Now you fetch the comb. Wrap it in a moist cloth to protect the larvae. Then you find the larvae which are going to be grafted. If you have a low power stereo microscope (10 to 30 x magnification) it is an advantage to use it, especially if you have not grafted before. Then you can see the larvae clearly and measure their size.

The larva should not be more than 2 mm long, you use the grafting needle as a ruler. You can use the youngest larvae for grafting, and many people prefer that. You can also use the larvae which have gone through their first molt and become so big that they are easy to take. If you use the microscope and the ruler you learn fast how to recognize larvae of the right size.

By grafting you place the needle below the larva and lift it together with a little of the fluid food on which it is laying (fig. 8). You move the larva to a cell cup and place it on the bottom. The larva must not touch anything at all because it is easily damaged. Experiments have shown that it is unimportant which side of the larva is up. Don't bother if you turn the larva upside down during the grafting.

You can place the larva on a small drop of royal jelly thinned with water 1:1. You can use the "wrong" end of the needle for that transfer, or a match. This is called *wet grafting*. The advantages

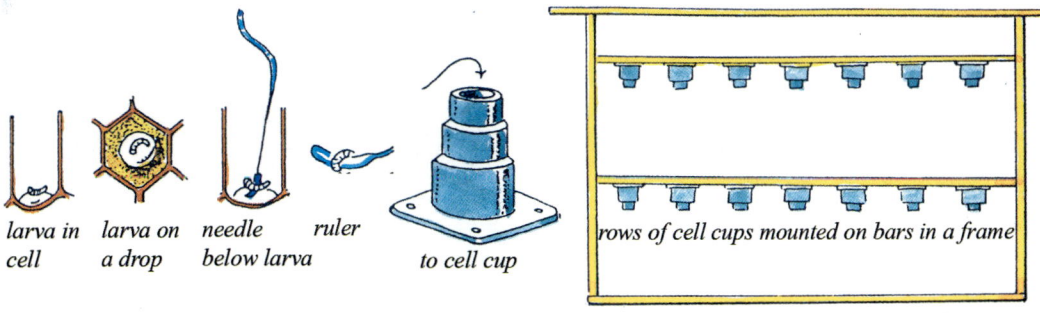

Fig. 8. Grafting, scheme.

1. PRODUCTION OF QUEENS   11

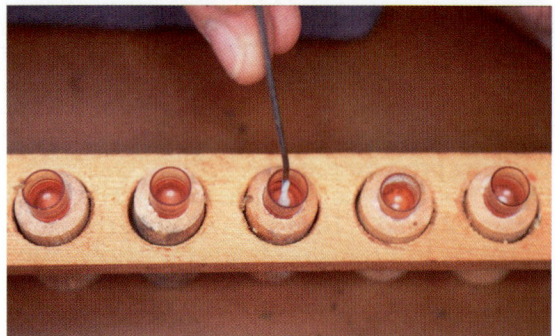

Fig. 9. Cell cups get a drop of royal jelly before grafting.

Fig. 10. The queen cells are finished. The bees nurse the larvae.

is that the larva slides easily off the needle and that you can wait longer time for transferring of the larvae (fig. 9). However, the liquid is not necessary. Without it you make *dry grafting*.

Transfer to the cell builder

The frame with larvae in cell cups are transferred to the cell builder so fast as possible, latest 3 hours after grafting. By dry grafting the larvae should be transferred nearly instantly. If you need more time *wet grafting* is a must. The frame is put into the empty place in the hive and the following day you control if the larvae have been accepted (fig. 10 to 13). That is the case when the larvae are fed and the building of queen cells around the larvae has begun. Badly attended larvae are discarded.

The queen larvae use 5½ days to grow up in the open cell. Then it is sealed with wax by builder bees.

4.5 to 5 days after the grafting every queen cell is placed in a cage (fig. 14, 17). Before the caging you control the cells which should be well finished. However, it does not matter if it is lopsided.

You could wait with the placing of the cage until the 10th day after the grafting but there is a risk that the bees fill the spaces between the queen cells with honey cells. The queen leaves her cell on the 11th or 12th day after the grafting. While she is inside her cell she might be sensitive to pressure which deforms the wax of the cell. This can kill the queen.

Incubation

While the queens are caged you can treat them in two ways:

1. *The queens stay inside the cell builder* and run out inside their cage 11 to 12 days after grafting. The cages must be taken away very soon because the

Fig. 11. 22 larvae have been grafted, 17 of them are accepted. The sealed queen cells have different shapes.

Fig. 12. Something has gone wrong. One queen came out first and she killed the other queens.

Fig. 13. The collector bees have been busy. In need of room the bees have built honey cells among the queen cells.

Fig. 14. Queen cells are placed in separate cages before going to the incubator.

Fig. 15. Row of queen cages inside the incubator.

Fig. 16. Incubator with the cages of fig. 17. (Swienty).

workers sometimes bite the feet of the queens and thus damage them. These queens will not be accepted.

2. *The incubator.* The cages with the sealed queens can be placed in an incubator as soon as they are sealed, or you could wait until the 10th day after grafting (fig. 15). The queens leave their cells in the incubator. They bite the lid off from within, press it down and creep out (fig. 19).

The incubator is an insulated box (fig. 16). A thermostate keeps the temperature at 34,5^0. You have to control the temperature with a thermometer. The relative humidity must remain between 40 and 60%. It is done by placing one or more flat bowls with water on the bottom of the incubator. The humidity is controlled by a hygrometer. Supplying humidity is provided with a water atomizer. You can buy an automatic humidity control.

You can buy the incubator or you can make it

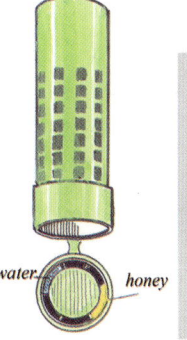

Fig. 17. Plastic queen cage with water and honey in the lid.

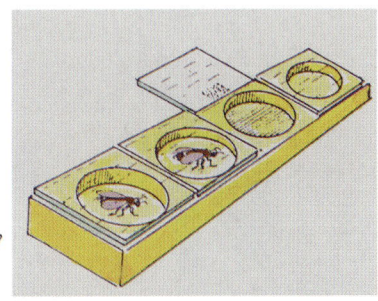

Fig. 18. Row of containers with separate glass lids for queens awaiting their placing in a transport cage.

1. PRODUCTION OF QUEENS 13

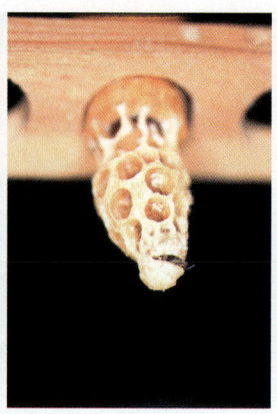

Fig. 19. The queen emerges from her cell. First the tongue goes out after honey. Then it bites through the wax and opens the cell. At last it emerges completely

yourself, for example in an old refrigerator.

The queens must have food as soon as they have left their cells. A drop of honey so big as the head of a match is used as food (fig. 17). 6 hours later they must be fed again with a drop of honey water on the net of the cage. They can live in the cage for a week if they get food two or three times a day. Usually, they are placed in a transport cage few hours after their emergence together with from 6 to 10 attendant bees, 4 to 6 days old. You could take them from a pollen comb in a queen less cell builder.

How to obtain young bees

Young bees for many transport cages can be taken from a super with unsealed brood. The super should be separated from the box which contains the queens of the colony by a queen excluder. So you are sure not to take the queen with you.

Fig. 20. Left: Kicking the box with bees. The old bees leave. Right: Water atomizer for spraying water on the bees.

Young bees must also be used in the nucs where the young queens are going to be placed. You have to use 2,5 decilitres of bees for each nuc.

The tool you need is a super with a bottom or a cardboard box of approximately the same size (fig. 20). A number of combs with brood are shaken above the box and the bees fall down into it. Now you kick the box several times and the bees fall down on the bottom. The pollen and honey collectors which are inside the box fly away and return to their own hive. The young bees stay in the box because they never have been out of their hive and do not know where they live. The young bees are now drawn from the box by a container made from the lower part of a cardboard milk container. It contains the 2.5 to 3 decilitres of bees which are needed for a nuc.

If the bees are going to be used on a mating station they must be strained for drones (fig. 45).

Marking of queens

Queens ought to be marked according to the international colour code system (fig. 21). Then a queen will be easy to find inside the colony and you can see her age. The colour is dissolved in alcohol or cellulose thinner when you buy it. The colour is placed on the scutellum (fig. 21) with a pinhead. The pin is fixed in a cork or a piece of wood. The solvent evaporates in a couple of minutes. You can buy a marking pen with the colour of the year.

It is important that the colour is placed only on the scutellum. If some colour hits the wings it could influence the mating flight badly. If a little colour goes between the head and the thorax the queen cannot move her head. She cannot survive that.

14

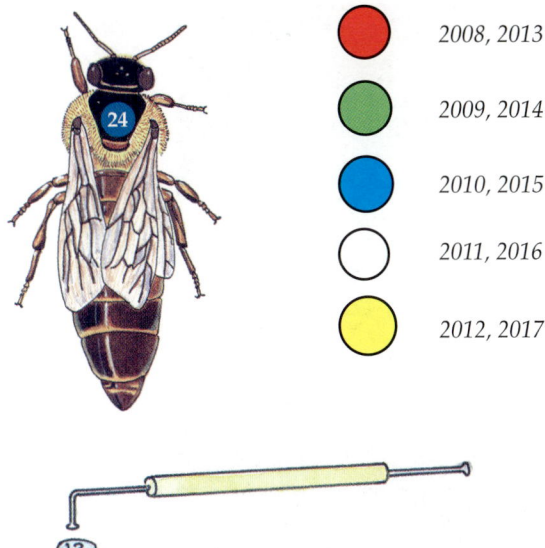

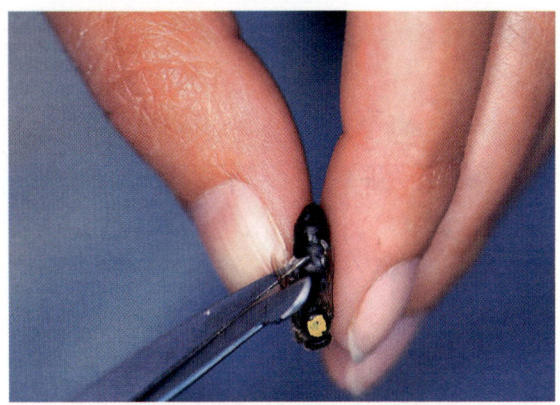

Fig. 23 a. Cutting the wings of a queen.

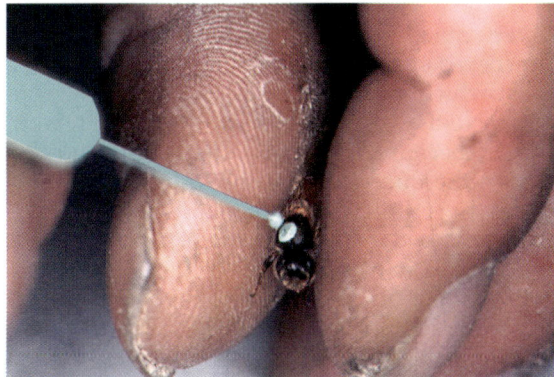

Fig. 21. Left: Queen with a number on her scutellum. Right: Colours of the years for marking of queens. Bottom: Tool for placing a number platelet on the scutellum.

You fix the queen between the thumb, the index and the middle finger while she is marked. Her abdomen must never come under pressure. You can practice on drones.

You can mark the queens with plates which carry the number 1 to 99 on a background of the colour of the year. The plates are delivered with glue and a tool for placing plates and glue on the scutellum. These plates are used to secure that you know exactly which queens you have. This is important for breeders and in genetic research.

The tool for marking is a wooden shaft with a pin in each end (fig. 21, 22). One of them is straight, the other bent in a right angle. Note:

Fig. 22. Marking of a queen with colour.

A steel pin can only be bent after glowing. The head of the straight needle is used to transfer a colourless glue (lacquer for fingernails dissolved in acetone) to the posterior part of scutellum. After that you moisten the head of the bent needle with your tongue and picks up the plate for fixing on the queen.

The numbered plates can be placed so as to be read from the side of the head, from the abdomen, or from the left or the right side. Then you can decide that numbers which you read from the abdomen means 1 to 99, read from the left side 101 to 199 etc. You should not use numbers which can be read otherwise when they are seen upside down: 6, 9, 66, 68, 69, 86, 89, 96, 98, 99.

Poul Erik Sørensen mark queens of the same stock, for instance sister queens, with two colours. First a drop of colour transferred by a match, then a smaller drop of another colour transferred with a pinhead. The 5 colours can be combined in 32 ways. In this case you can see if a queen has arrived in a wrong colony or if there is a failure in the notes.

When should you mark the queen? If she has lived in a small colony throughout the winter you can mark her in springtime and at the same time cut her wings. After that she is returned to her colony without problems (fig. 22, 23a and b).

The cutting of the outer half of either the right or left pair of wings has the advantage that the queen falls to the ground if she leaves the hive with a swarm. After that the bees return without her. If you see a queen with unclipped wings in the hive she has been exchanged by supersedure.

1. PRODUCTION OF QUEENS   15

Fig. 23 b. Queen with cut wings and marked with two colours on the scutellum.

Transport cage and candy

The transport cage is used during shipping of the queen with her attendant bees and for introducing the queen in a colony.

Today a cheap and effective cage of plastic is used (fig. 24). Its bottom plate is taken off while it gets its load of candy. The bottom plate can be pushed along rails. That is done during the loading of the cage with the queen and her attendants.

The part which contains the candy is closed by a lid during transport. The lid is broken off by the introduction in a colony and now the bees can eat their way to the queen. Holes in the cage in the part with the bees is necessary for their respiration. During the introduction the bees can get in touch with the queen through the holes.

By the introduction you place the cage between two brood combs fixed by a match through the eye of the cage. Before the introduction you open the cage in a closed room with windows. The attendants and the queen fly towards the light. You catch the queen and put her in the cage. In this way you get rid of the attendants which could bring diseases and parasites with them.

Queen candy is made by invert sugar mixed with powdered sugar. It must not dry out and form a crust which makes it impossible for the bees to eat it. Neither should it suck water vapour from the air because it then becomes fluid and can drown the queen and the bees. The production of candy is difficult and demands much experience. It is better to buy manufactured candy. It can be used in mating cages.

When you put bees into the cage you start by placing 6 or 10 attendants and then the queen. You can anaesthetize the attendants in a bucket filled with CO_2 from a steel flask. CO_2 is heavier than air and stay in the bucket.

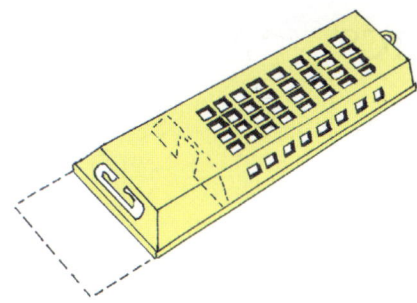

Fig. 24. Cage for the queen and 6 to 10 attendants. It is made of plastic 80x36x13 millimetres. The bottom plate can be moved (dashed line). The left part of the cage is for candy. The split in the end plate of the candy chamber is broken off when the queen shall be eaten out. Place the cage with the bottom plate up, then sugar grains drop out of the cage.

Shipping
The queens can be sent by mail in solid envelopes with small holes for the respiration of the bees. You write "Live bees" on the envelope. The receiver has to know when the queens arrive. Then he can make preparations for their introduction.

Much transport of queens is done by air. For instance Hawaii queens are flown to Denmark, that is 12.000 kilometres across the North Pole. Both mated and unmated queens are sent in that manner.

2. Mating nucs

Unmated queens are ready to mate when they are 6 or 7 days old. Bad weather can delay the mating a month. During that time the queens can mate without problems.

Queens must be attended by a colony before and after the mating. It can be done in a small colony but usually a nucleus box or "nuc" is used. A nuc only needs 2½ decilitres of bees and therefore it is easy to get enough of bees.

Many different nucleus boxes exist. Here two types with 3 or 4 combs are described (fig. 25, 26). The one comb nuc is described in chapter 4. The nucleus box with 3 or 4 frames can be used by beekeepers which mate their own queens without control of the mating. The one comb nuc (fig. 44) is used on mating stations because you can control if it contains drones.

The nucleus box with 3 or 4 combs is made by polyurethane foam or similar materials. The material must isolate well. If that is not the case the small number of bees cannot maintain the temperature they need. If the temperature rises too much the bees leave the nuc box. If the temperature is too low the development of eggs and larvae could stop.

The nuc has a room for 750 gram of candy and another for the bees and the queen. There must always be candy inside. In the beginning it is needed because none of the bees are old enough to collect food from the flowers. When the bees have got the age for collection they are too few to fetch enough of honey and pollen. You can cover the candy partly with a piece of greaseproof paper to prevent the bees from being covered with

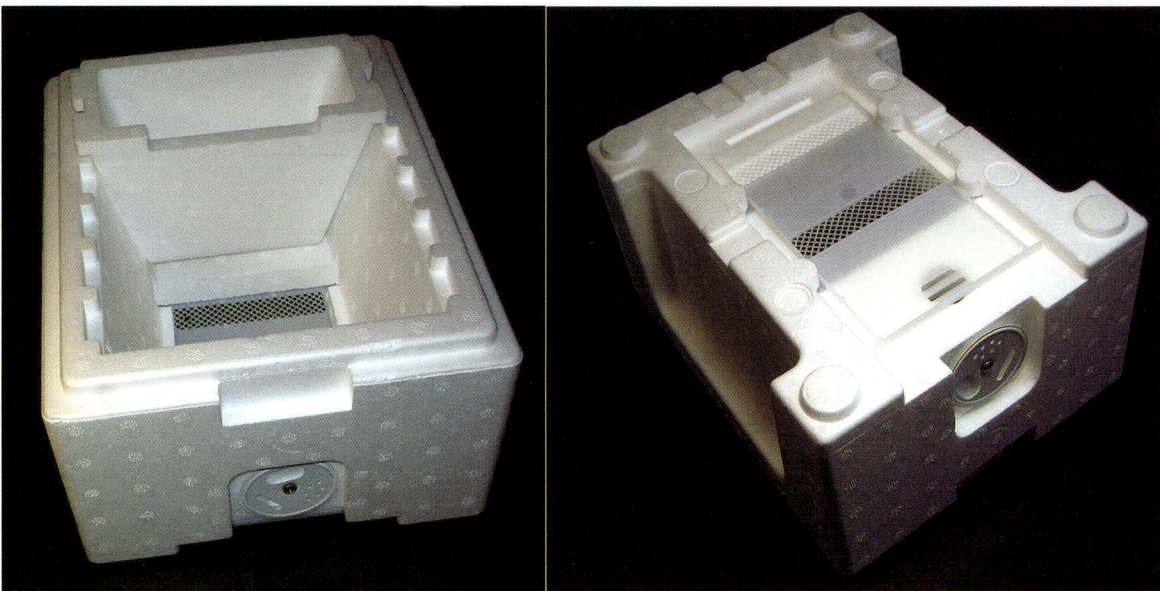

Fig. 25. The Kieler nuc from above (left) and from below (right). It has four bars with slits for foundation and a room for candy which can be taken out and cleaned. The plate of the bottom can be placed such that the bees are enclosed but get air through two grills. The opening in the front is covered by a circular plate which can be turned so that the queen and the workers can fly out, or only the workers can get out, or only the air can pass through it.

candy. If needed, the bees can penetrate the paper.

In the room for the bees there are 4 bars with a slit on the lower side. Here you place a piece of foundation 2 or 3 centimetres high and nearly so long as the bar. Before that, dip the foundation in a container with a little melted wax, 0,5 cm deep.

The bees build combs beginning at the foundation. The sides of the Kieler nuc box are sloping and therefore the bees rarely fix the comb to the walls. There are no steel wires in the comb. Because of that you never should hold the combs horizontal during the inspection, in stead you ought to hold them in a vertical position. If you hold them horizontal they could break because of their own weight.

The interior is covered with a piece of plastic fixed with two thumbtags. Then you can inspect the bees without disturbing them just by opening the lid.

In the bottom of the nuc box there are holes for ventilation. A movable plate can close the entrance while the ventilation holes are covered by a grill. This is the position of the plate during transport. The plate can also be positioned so that the entrance either is covered by a queen excluder or is uncovered for the mating flights of the queen.

The nuc is covered by a lid. It is kept in place by a stone or by rubber bands.

How to use the nuc
You can fill the nuc in this manner: Place the candy in its room and make the wooden bar with foundation ready. Now the queen is placed on the bottom and you see if she looks healthy and walks well. If she walks upwards on the walls then spray her with a little water. Note: The queen should not be placed in the nuc during her time of flying which is about 11 to 17 o'clock because she could fly away. However, if she is less than two days old she will not fly.

When the queen is accepted by the beekeeper 2½ decilitres of young bees are poured into the nuc. Do it in the evening when the bees are most quiet. After the 1st of July you use 4 decilitres of bees because they can defend their nuc against robbing. You can make the bees quiet by spraying them with water before you pour them into

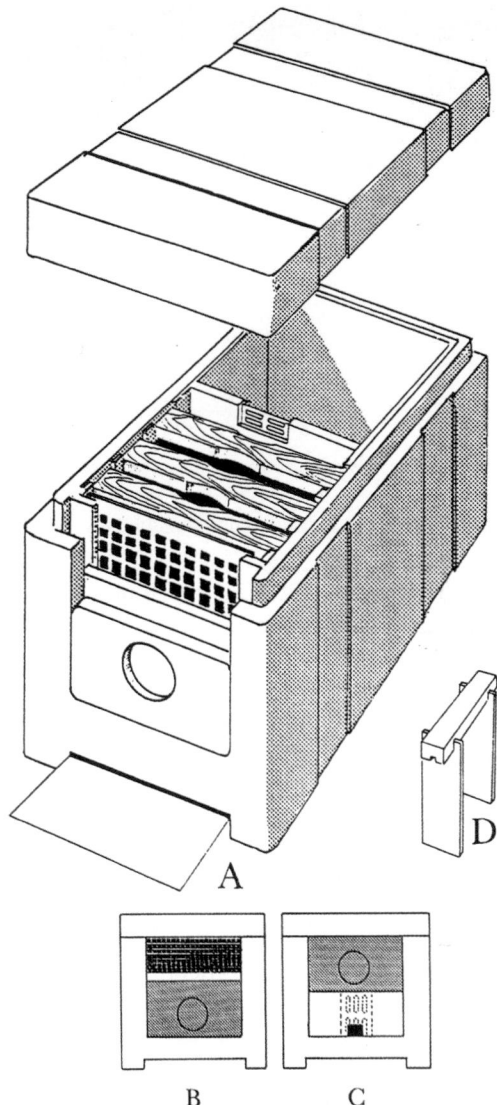

Fig. 26. Swienty's nuc (Swi-Bi) has vertical sides and the lid can be pressed in place. It can be further secured by rubber bands in the slits. Because of that this nuc is easily moved. The bottom plate has no holes, when it is drawn out you see a big opening which is used for filling bees inside. There are three frames for combs with wood on three of the sides (D).

The entrance to the candy is controlled by a queen excluder. Therefore, she cannot enter and drown in the candy.

In the front the nuc has a movable plate. In the lower position the air goes in through the grill (B). When the plate is pushed upwards you see the entrance (C). When the queen must stay inside a piece of queen excluder is placed in front of the entrance (C, dotted). Usually, you will not push the plate to its uppermost position because it then will block the ventilation grill.

Fig. 27. Nucs placed in groups of four. The queens and bees find their own nuc because of the colour of the lids and because the entrances are in different directions.

Fig.28. Open nuc. The bees have built honey cells above the candy. New candy must be added. A queen is added in a cage.

the nuc. This method is safe. After that the nuc is covered with transparent plastic.

You can also add the queen in a transport cage without attendants (fig. 28). There should not be more than 2 millimetres of candy because the few bees cannot eat through a thick layer. This method is particularly necessary if the queen has arrived with attendants. If she is placed in the nuc without protection she could be killed by the bees, probably because she smells wrong, i.e. of the attendants.

Now close the nuc with the bottom plates so that the holes for ventilation are covered by a grill. If

Fig. 29. The combs of the nuc are soft and must never be held in a horizontal position because they easily could break. In stead, you turn them keeping them vertical and place them in the lid.

the transport will last a long time you add water by spraying it through the grill. The entrance is opened when the nuc is placed on the mating place. As the queen fly out to mate when she is 6 or 7 days old you can wait 5 or 6 days before you move the nuc to its final place. The nuc should be placed there in the evening. Thus the bees have time to accustom themselves to the place. The nuc is placed in half shadow. If the queens are going to be mated somewhere around your bee yard the nucs are placed at some distance from it to lessen the risk of robbery.

The nucs are placed so that the entrances turn in different directions. This makes it easier for the bees to find their own hives. You could further help their orientation by painting the lids in different colours (fig. 27).

After one or two days the queen flies out to orient herself. Then follows the first mating flight. Two or three flights on successive days suffice to get the mating she needs. On the average she mates with 17 drones. If you disturb the colony while the queens are flying the bees could hinder her return or cling around her. Therefore: Stay away from the nucs in her flying time.

You don't need to inspect the nuc before the queen is 12 or 14 days old. If the weather has been good she could be laying eggs. She stays in the nuc until there are larvae and sealed brood (fig. 30). The queen is given to a colony, se page 31.

The bees of the nuc can receive a new queen

after having been queenless 4 or 5 days. But they need about 100 young bees. Pour them in front of the nuc and they move in. Before the queen is transferred in a transport cage you inspect the combs for emergency cells. They are taken away. You only need to add bees when you place the second queen. When you place the third queen there are a sufficient number of workers in the nuc.

Continous management of a nuc
A nuc can be used for the production of 4 queens one after another. The first queen is placed in the nuc as described above. When the queen is 18 days old she is removed and a queen cell is placed in the nuc two days before emergence. She is accepted as soon as she emerges.

If there are many nucs on the mating station you change the queens with an interval of 20 days. A row of queen cells are taken from the incubator and are transported, wrapped in a thick piece of cloth. The queen cells tolerate 12 hours of transport and survive becoming wet or cooled, but then they emerge later.

By the first changing of queens you remove two brood combs and add two bars with foundation. The queen cell is placed between two brood combs. 350 to 400 grams of candy is placed in the room for food. At last, 100 young bees are placed in front of the entrance. They move in.

If the changing is made after the 15th of July you don't remove brood combs because there is a sufficient number of young bees.

By the second changing you add from 200 to 500 grammes of food. It is not necessary to add bees or brood combs.

By the third changing you add 400 to 500 grammes of food because the honey flow is diminishing. This changing should not be done later than the beginning of August because the probabilities of mating is reduced.

Queens which have not mated before 20 days have passed are killed.

The queens which have emerged in the nuc have to be marked (fig. 21, 22). She is removed from the nuc in a cage. In another cage you keep 8 to 10 attendants from her nuc. The queen is marked, for instance inside the car, and one or two minutes later she is placed in the cage with her own attendants.

Fig. 30. Comb from a nuc. The sealed brood shows that the mating has been a success.

Clearing after the mating season
In the beginning of September you make a colony of the bees from 30 to 50 nucs. Shake the bees down in a hive with 10 combs (4 pollen combs and 6 empty combs with cells. Then you feed for winter. If a nuc has been queenless for more than 12 days a queen has emerged, and she will be among the bees. Some days after the colony has been made you inspect it and the queen is taken away, if she is there. A new marked queen is transferred to the colony.

Brood combs from the nucs are placed in a super which is half as high as a normal super. You divide it and thus there is room for the combs. From there the bees walk through a grill to a normal super with honeycombs and a queen inside a cage. In this way you make a new colony.

The nucs and their bars are cleaned in hot water with a neutral detergent for washing up. Transport cages can be placed in a sac and washed in a washing machine by 60^0. Use a detergent for a dishwasher. The bottom plates don't tolerate cooking.

3. Alternative methods

In this chapter methods for large-scale queen production and one method for small scale queen production are described.

1. The queen production and breeding of Poul Erik Sørensen
Poul Erik Sørensen produces thousands of queens every year. Grafted queen larvae from selected breeding colonies are placed in a *starter* colony where they are accepted and are fed by the bees during 20 to 24 hours. After that, they are moved to a *cell builder (finisher)* where they continue their growth until sealing. This part of the process lasts 4 days. Now the sealed queens are transferred to an incubator where they emerge from their cells. Then they are placed in a transport cage with 6 to 10 attendants. They can either be sold as *unmated* queens, or be placed in nucs

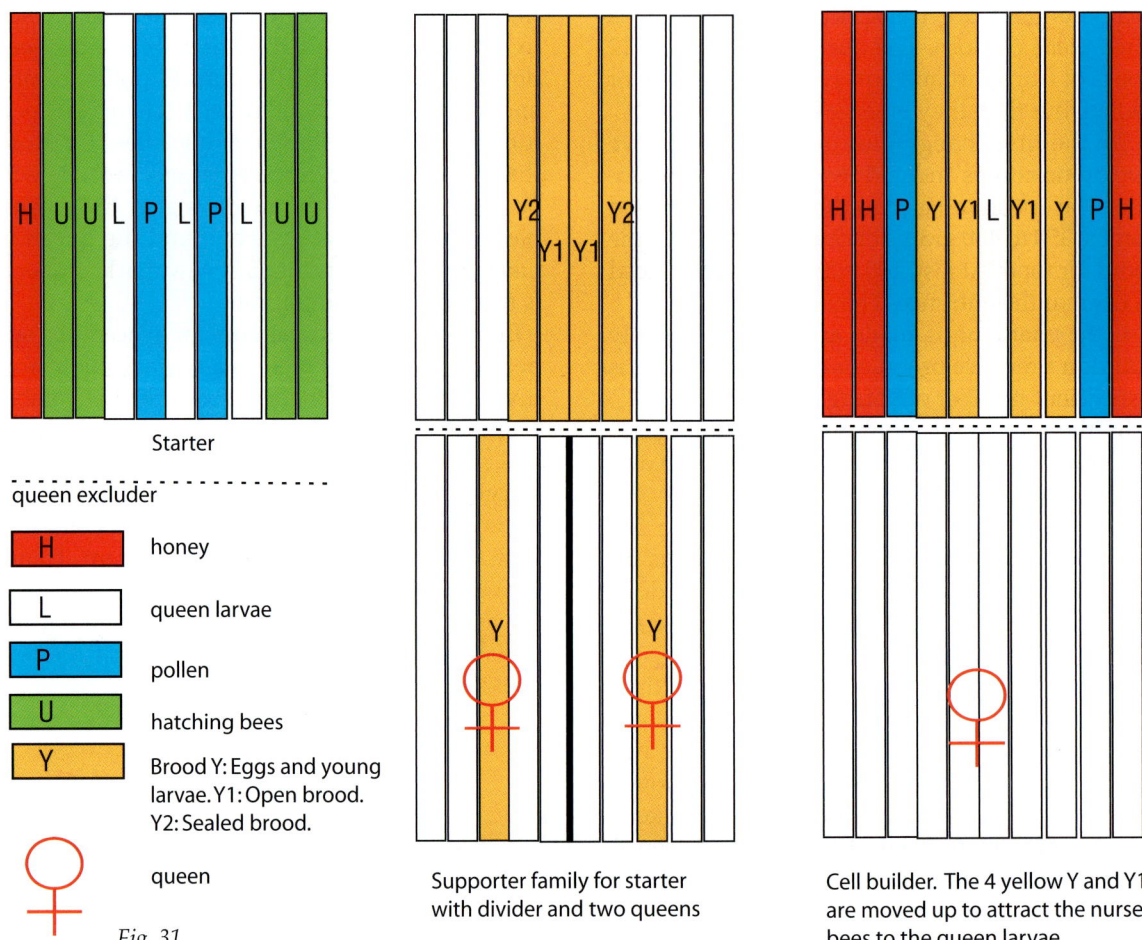

Fig. 31

Supporter family for starter with divider and two queens

Cell builder. The 4 yellow Y and Y1 are moved up to attract the nurse bees to the queen larvae

and mated in an isolated area in Jutland and sold as *mated queens*. Queens for breeding are mated on the island of Tunø.

The first grafting is made on the 20th of May. New graftings are made every week until the 20th of July. This can only be done by adding brood combs to the starters and cell builders after a system described below. All of the colonies are fed with sugar water 1:1 because they should feel a continuous honey flow.

The starter is a fast growing colony (fig. 31). It is made queenless 9 days before the colony is going to be used. The bees build emergency cells. The cause of the 9 days is that it lasts 9 days from the last egg has been laid and to the sealing of the last emergency cell. Before the grafted larvae are added the emergency cells are removed.

From 60 to 80 queen larvae are placed in three frames (fig. 32). They stay in the colony between 20 and 24 hours. Then they are distributed to three cell builders. Few hours later 3 new frames with grafted larvae are added (fig. 33).

The bees of the starter must be in the nurse stage. Every week 4 combs with emerging bees are added, they come from two support colonies, one on each side of the starter. The bees of the starter get two pollen combs and one honey comb. They are also fed with sugar water as described above.

Two supporter colonies to the starter. Each of the two supporters has two storeys separated by a queen excluder. The lower storey is divided in two by a plate. In each compartment there is a queen and 5 combs (fig. 31). Every week a comb with foundation is placed in each of the compartments. It is immediately covered with cells and filled with eggs. One week later it is moved to the super, where the larvae are sealed and pupate. A new comb with foundation is placed in the two compartments of the lower storey.

Three weeks after the supporter has begun to work there is one comb with young larvae in each compartment, while there are two combs with newly sealed larvae, and two combs with sealed bees which are ready to emerge.

From each of the two supporter colonies the starter gets two combs with emerging bees, in all 4 combs every week. Thus the starter always will

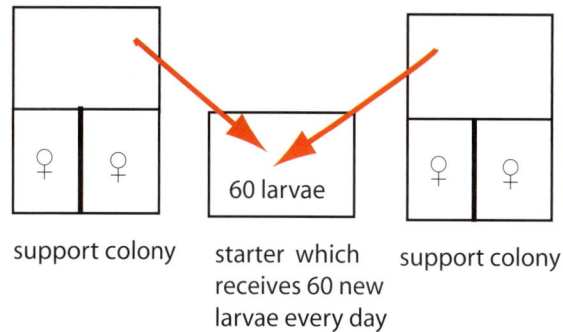

Fig. 32

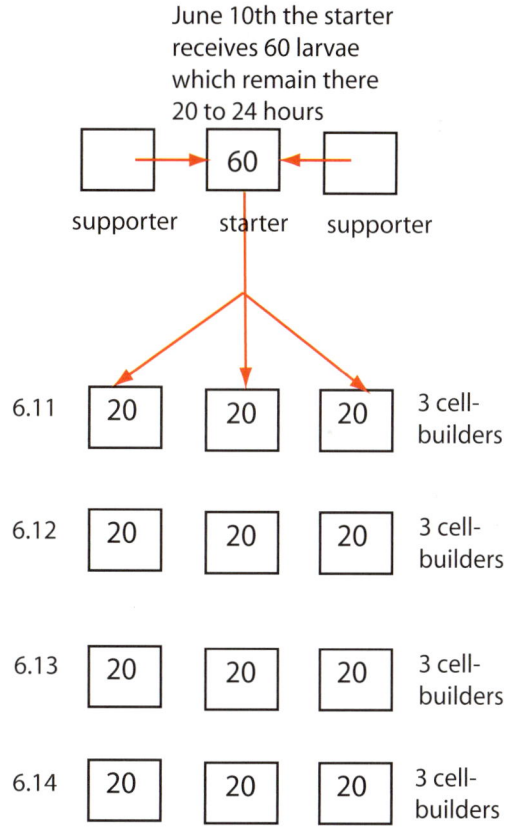

June 11th the 3 cell builders get 60 queen larvae from the starter. They remain here for four days. June 12th the next 3 cell builders get 60 queen larvae from the starter. The same happens June 13th and 14th.

A production line consists accordingly of 1 starter + 2 supporters + 12 cell builders = 15 colonies.

Fig. 33

3. ALTERNATIVE METHODS 23

have a sufficient number of nurse bees.

Superfluous collector bees from the starter go to the supporters. The supporters get ½ to 2 litres of sugar water 1:1 every day. They function from May 20th to July 20th when the queens are produced.

The finisher is a colony in two storeys (fig. 31). In the lower storey is a queen and 10 combs. The queen larvae are nursed in the super above a queen excluder. The finisher is only able to nurse 20 queen larvae and therefore you need three finishers for 60 larvae from the starter.

The queen larvae stay 4 days in the finisher. Then they are sealed and transferred to the incubator. A new set of queen larvae are transferred to the finisher. They are placed in the super, where two pollen combs and three honey combs are placed, too. Every day they are fed with ½ to 2 litres of sugar water 1.1. The finishers function through all of the queen rearing season.

Computation. One starter produces 60 queens every day, and every day 60 queens are distributed to 3 finishers where they stay 4 days. 12 finishers are needed to receive the queen larvae from 4 days (fig. 33).

One starter demands two supporters and 12 finishers. A full production line demands accordingly 1+2+12= 15 hives.

The production cannot be exactly the same every day because bees are living things and they do not behave in the same way. Environmental factors influence the production, too. Poul Erik Sørensen is satisfied with a production success of 80%.

2. Queen right cell builder

You could use a queen right cell builder throughout the season. The method is used by big queen producers. The cell builder has two boxes divided by a queen excluder and a space wherein a thin metal plate can be inserted. Thus the bees of the two boxes can be separated completely. The queen lives in the bottom box. The super has its own entrance. The cell builder works in this way (fig. 34):

    A. Queen right cell builder with an entrance in each storey. Both are open all the time.

B. A short time before the grafting the two storeys are separated by a metal plate. Soon the bees of the super feel queenless. You add a comb with larvae + bees and combs with emerging bees to be sure that the grafted larvae are nursed well.
C. Frames with grafted larvae are added.
D. 24 hours after the arrival of the grafted larvae the metal plate is taken away.

5 days after the grafted larvae arrived the sealed queen cells are placed in the incubator. Then you have to come back to *start* and can add grafted larvae again. This can be done every week throughout the season.

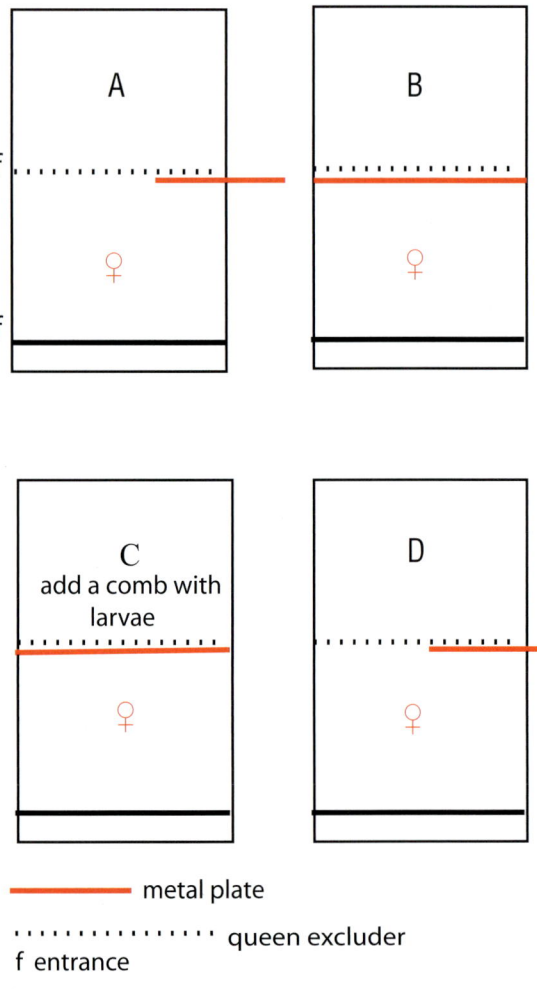

Fig. 34

3. The starting box method

This method is used to start up the growth of the queen larvae so that they are easily accepted inside the cell builder (fig. 35, 36). You should use.
1. A box with a bottom of wire net and room for 6 combs.
2. A plate which divides the box in two. The plate must be higher than the box.
3. Two lids which together cover the box.
4. Three combs with bees. The combs should contain pollen and honey but no eggs or brood. Young bees from two brood combs.
5. A cool and dark room.
6. A frame with 10 or 12 grafted larvae.

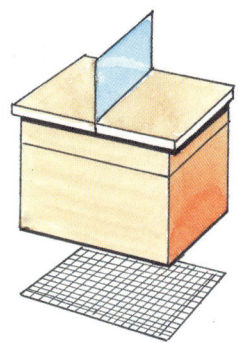

Fig. 35

The starting box is used in this way: A box with room for 6 combs is divided by a plate which separates the two parts completely (fig. 35).

A lid is placed over one of the compartments. Three combs with honey, pollen and many bees are placed in the other compartment. Young bees from one or two combs with young larvae are added. The bees are covered with a lid. The box is placed in a dry and cool place such that fresh air can get in through the bottom (fig. 36).

The bees will soon discover their desperate state. They have neither a queen, nor small larvae which could be used in emergency cells. They move around sounding in the way queenless bees do. Don't open the lid, the bees will escape.

4 hours later a frame with 10 or 12 grafted larvae are placed in the empty part of the box. Put on the lid. The dividing wall is taken away and the queenless bees accept the larvae, start the building of the queen cells, and feed the larvae. In a minute or so the bees become quiet.

The bees are standing 24 hours in a cool and dark place. After that the queen larvae are fully accepted. The combs are returned to a super of their original hive together with the queen larvae which are nursed here. The bees nurse the queen larvae in the super even if there is a queen inside the lower box beneath a queen excluder. In fact, you can place grafted and accepted queen larvae to whatever queenright colony you wish if you separate the larvae and the queen by a queen excluder.

24 hours before the young queens open their cells they are placed either in an incubator or in cages within the colony.

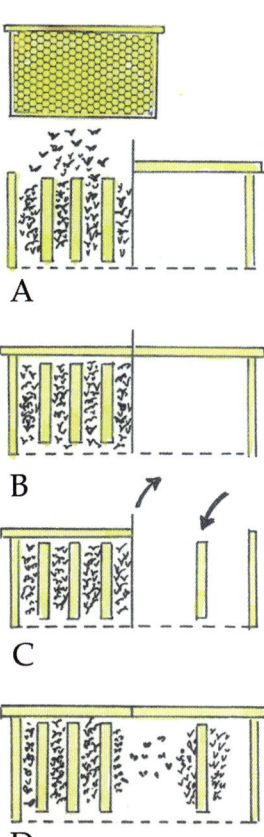

Fig. 36

The nurse colony must be strong. As a rule, you should not exceed the maximum of 10 or 12 queens to be sure that the bees can feed them.

4. Mating stations

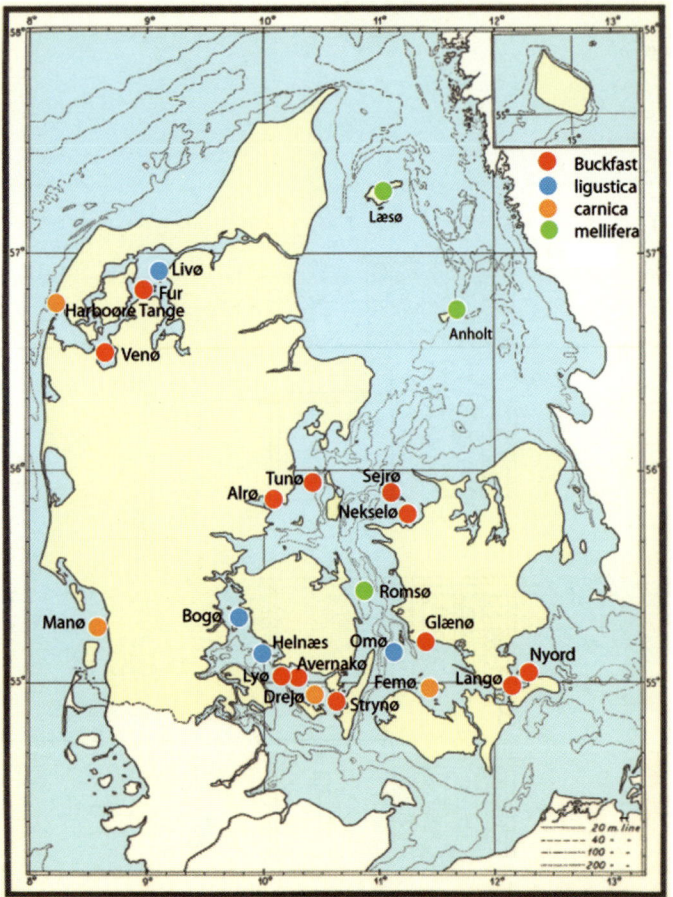

Fig. 37. Mating stations of Denmark 2007.

The queens fly out to mate in a drone congregation area where drones from many bee yards come. There she is mated. That means that the beekeeper knows nothing of the descendence of the sperm. He cannot answer the crucial question of breeding: Who is the father?

It is a fact that the genes from the mother and the father are of equal importance for the offspring. If you wish to breed good bees you cannot allow your queens to mate after their own choice. The problem of how to control the mating is solved in two ways: Mating on a station or by instrumental insemination. Both methods demand much work and can best be done through collaboration between many beekeepers.

Nearly all of the isolated mating stations in Denmark are situated on islands because drones and other bees don't like to cross the water. If an island is placed 3 kilometres from the nearest land it is safe (fig. 38).

The distance has been found through experiments. Mutant bees with the leather brown colour *cordovan* which is easy to recognize were placed at islands in different distances from the nearest land. When cordovan mate with cordovan all of the offspring becomes cordovan. If cordovan mates with yellow *ligustica* bees or black *mellifera* the workers in the offspring are yellow or black. It is thus easy to see if foreign drones have visited the island. Cordovan bees have two recessive genes for the colour cordovan. If the cordovan gene is together with either the yellow *ligustica* gene or the black *mellifera* gene then they dominate the cordovan gene.

Where isolated islands are unavailable the mating stations are placed in isolated mountain regions or big forests, far away from other colonies.

Organization of the queen breeding

The Danish mating stations (fig. 37) are maintained by associations of queen breeders or private breeders. Every beekeeper can send his queens to the mating stations of the association *Dronningavlerforeningen af 1921* (The Queen breeders' association). The payment is moderate.

A mating station cannot be maintained unless the breeders have well defined aims for the use of it, for example breed better *ligustica* bees, or maintaining the characters of certain stocks of

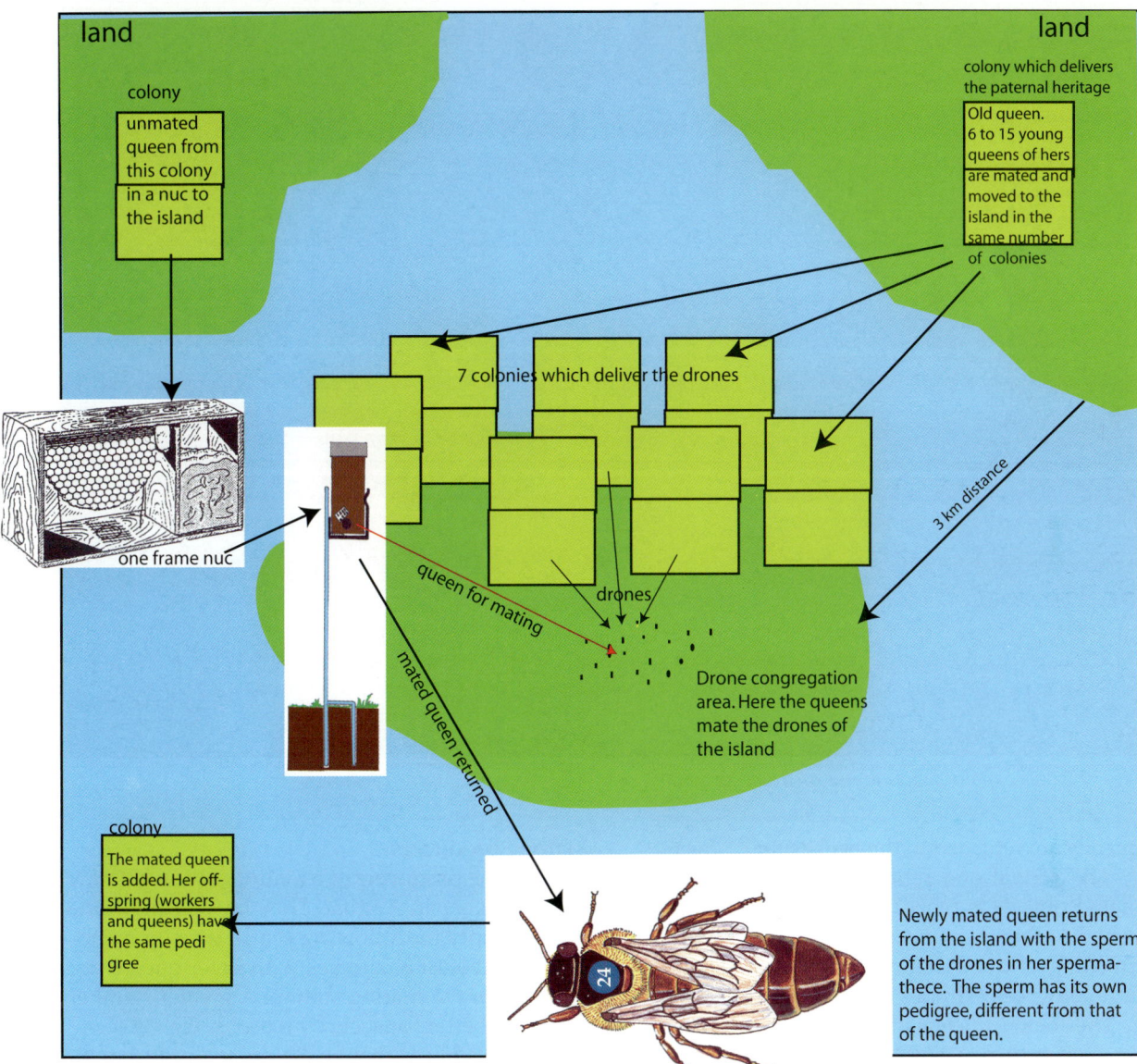

Fig. 38. Mating on an island, schematic. Only 7 drone colonies are drawn, but more colonies are better.

Buckfast bees. This demands a meticulous maintenance of records and pedigrees, se chapters 6 and 7.

If an association or a private beekeeper wants a mating station they must apply for it in a scheme which is sent to a governmental office. The permission is valid for 5 years and is given on certain conditions, for instance must the colonies of the mating stations be controlled for certain diseases, and the beekeeper of the station is obliged to report annually about the work which has been done. All the colonies must have queens of the same descent, and queens of other descent must not be introduced.

The idea of the rules is that the matings should be controlled and safe. This goal is attained by using the colonies of the station as drone producers. Their genetic uniformity is obtained by using sister queens in all of the colonies.

In June, July and the beginning of August unmated queens are sent to the mating stations in one-comb nucs. Two weeks later the mated

Fig. 39. These colonies deliver drones on a mating station.

queens are returned to the beekeeper.

The arrival and return of the unmated and mated queens is scheduled. The dates are published so that every beekeeper can plan his own breeding of queens and their mating. The race and pedigree of drones is published, too.

The mated queens are placed in colonies after their return. Many beekeepers keep records of them and will often choose their offspring for further breeding.

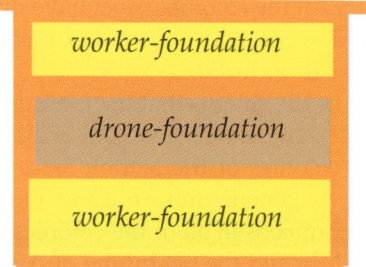

Fig. 40. Drone foundation between two parts of workers' foundation as used on a mating station.

Drone producers

The drone producers of a mating station get new sister queens in August. They lay eggs in drone cells in April the following year. It is recommended that at least 6 colonies with sister queens are placed on the mating station. A greater number is much better for security reasons. Fig. 39.

It is possible to change the drone producer colonies in the spring time. Before they are moved to the station they stand with a queen excluder at the entrance because alien drones must not go inside.

The drone producers must be very big and strong and have easy access to pollen and nectar. They should have room for food so that they can get sugar water in periods of lack of food. The hives are placed where the microclimate is best (high temperature, half shadow, little wind, easy access to water).

The drones produce only sufficient sperm if they have easy access to pollen and honey. Sometimes you see many drones in a colony, but

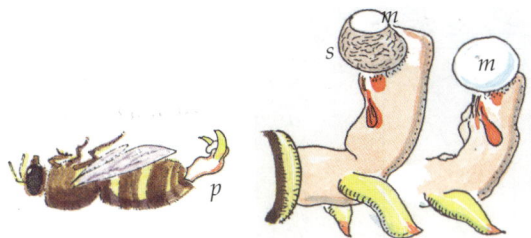

Fig. 41. Penis (p) is forced out. Sperm (s) and mucous (m).

when you examine them very few have sufficient sperm. That occurs at the end of the season, but it could happen, too, during periods of bad weather or prolonged drought. Then you must feed the colonies.

Control of sperm quality is done by pinching the drone (fig. 41). Then the penis comes out and you can see the brownish sperm on a white background of mucus. By examining a number of drones you can judge how much sperm they produce.

Some beekeepers use drone foundation to make the production of drones bigger. It can be done by placing it as a strip in the middle of a comb with ordinary foundation (fig. 40).

During prolonged bad weather conditions in July the drones can be thrown out, but not simultaneously in all the colonies. These drones can be saved if some of the drone producers are deprived of their queen. The homeless drones will go into these hives and are allowed to stay. To prevent that event all of the hives has a feeding container which is filled with sugar water by every visit.

The drones find their drone congregation area where the queens are mated.

Placing of the mating nucs

The drones demand a temperature of 19^0 C before they fly out. There must not be too much wind. The mating flight occurs from about 12 to 17 o'clock, summer time.

The queens are helped by placing the nucs in half shadow in places with little wind, for instance in a glade in a forest. The nucs can be placed on the ground, it is often done with Kieler nucs and similar types.

On mating stations where one-comb nucs are received you use a rack (fig. 42, 43). The racks carry insulated boxes for the nucs. The bees are

Fig. 42. Scaffold for a one-comb nuc to the right. The queen excluder has been turned away from the entrance. The drawing shows the construction of the scaffold. Left: Kieler nucs on the ground.

Fig. 43. Inspection of a one-comb nuc before it is placed in the insulated box on the scaffold.

4. MATING STATIONS

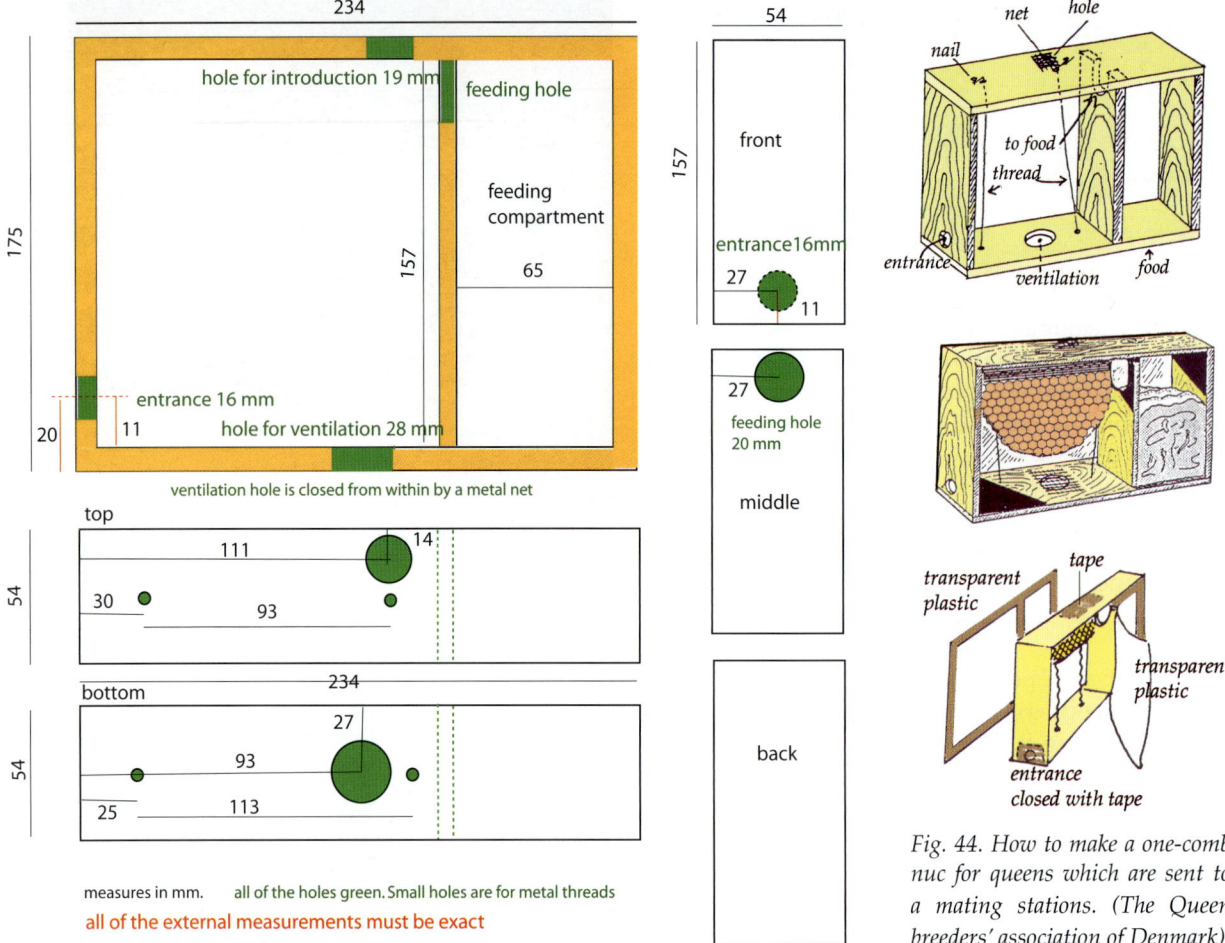

Fig. 44. How to make a one-comb nuc for queens which are sent to a mating stations. (The Queen breeders' association of Denmark).

helped to find their own nuc by turning the nucs so that their entrances turn in different directions, by painting the lids in different colours, and by placing the racks 2 or 3 meters from each other.

When the one-comb nucs arrive with bees and unmated queens they are placed in the insulated boxes. The entrance of the nuc is opened and it fits to the opening of the box, where the flight can be controlled by turning a piece of queen excluder.

The work on the mating station takes a long time. Because of that the number of visits is so small as possible. It means that the queens are brought home after two weeks on the station. Because of that the queens ought to be sexually mature, that is 6 or 7 days old when they arrive at the station.

If the weather permits about 80% of the queens will mate. But the condition is that every beekeeper fills his one-comb nuc correctly.

One-comb nuc

The queens are brought to the mating station in the one-comb nucs (fig. 44). The two greater walls must be of transparent plastic. This is demanded by all the mating stations because you can control if there are drones inside. If that is the case you lose the control of the matings. Therefore a nuc with drones is immediately returned to its owner. You can also see if the queen is there and if she has started the egg laying. The control can be done without opening of the nuc. The small size of the nuc makes it possible to transport many nucs at a time to the station.

The one-comb nuc is not insulated. When it

arrives at the station it is placed in an insulated box which belongs to the mating station. If you make the nuc yourself you must be careful to follow the measures of the drawing (fig.44). The nuc and the candy can be bought at the providers of outfit for bees.

The box has two rooms. The smaller one is for candy, the other one is for the bees. In the room for bees there are two metal wires which are used to fix a piece of foundation. Make waves on the wires, thus you are sure that the foundation stays in place.

Before you use the box you fix the plastic walls to the dividing wall with stales. One of the plastic walls is fixed with tape. The entrance is closed with tape, too, but not the ventilation openings. At last you fix a label with the number of the queen and the name of the owner.

Now the queen and exactly 2,5 decilitres of bees are filled into the nuc. After that the plastic wall is taped to the nuc. The nuc is placed in a dark and cool place. The optimal temperature is 15 to 18 degrees Celsius. The bees must have water, give them 2 or 3 squirts through the ventilation hole every day.

How to remove drones

How to get young bees for nucs is described on fig. 20. There must not be a single drone in the nuc which is sent to the mating station. The drones are removed in a Marburger box (fig. 45) with two empty combs with cells and some Apifonda sugar. On the lid is placed a plastic thread imbibed with queen pheromone as attractant. The young bees are placed in the open drawer and walk inside the box through a queen excluder of the plate type. If you use the thread type it will be filled with bodies of drones trying to go through.

Transport

The bees of the nuc build the comb with cells while it is in the house of the beekeeper. The ideal temperature is about 16^0 C. The advantage of the building is that the risk of absconding on the mating station is minimized.

The beekeeper carries the nucs to a centre from where they are transferred to the mating station the following day.

The nucs are placed in the car so that air can

Fig. 45. Marburger box. The bees are placed in the drawer and walk inside the dark box through the queen excluder attracted by a plastic thread with queen pheromone. The drones stay in the drawer. When they have been taken away you close the drawer and place the box in a cool place. Air goes through the net to the bees.

go through the ventilation openings. Before you go the bees get water from a spray. You keep the spray so that the bees can have more water during the transport if needed. In hot weather the bees must get water several times and you need to let fresh air pass the bees all the time.

The mated queens are returned to the centre in the same way as the unmated queens.

How to treat the mated queens

A queen which has been mated on a station is precious and ought to be treated in such a way that she is not killed when she is added to a colony. A safe method is to place the queen in a Kieler nuc or a box from a hive and add 2 or 3 decilitres of young bees. They accept the queen immediately. The small colony can be united with a queenless family.

You can also make a colony in a box with room for 2 or 3 brood combs with unsealed larvae and sealed pupae and 2 or 3 combs with honey and pollen. You add young bees from 2 or 3 combs. The colony is left for 6 or 7 days. The bees build emergency cells which are removed. At the same time you look after if there are enough of bees. If more are needed you shake bees off 1 or 2 brood

Fig. 46. Transport of one-comb nucs. They are placed in front of the mating station.

combs in front of the hive. The old bees fly home while the young bees march in through the entrance. Now you add the queen in a transport cage.

The colony can be united with a colony from which the queen has been taken away. The two colonies are placed such that they only are separated by a newspaper sheet. The bees bite through it and the colonies unite without problems.

The time of the year is important. In March and April the queen can be added without a cage. You just take the old queen away and place the new one on her place on the comb.

In the time when the number of bees grow fast and the honey flow is big you cannot add queens. You have to wait until after the first feeding.

In August, after the high season, the queen can be added without problems.

You should not add a queen while the honey flow is low unless you feed the bees.

While robbery is going on, or while the bees are preparing for swarming, it is impossible to add a queen. Sometimes there are two queens in a colony because of supersedure. It happens most often in August. You ought to control if there are two queens in the colony before you add a new one.

Unfortunately we know too little about the life rhythm of bees. It is decisive for the success when you add queens. Lack of knowledge about the life rhythm can result in loss of queens even for experienced beekeepers.

5. Instrumental insemination

By instrumental insemination sperm is transferred to the oviduct of the queen with an apparatus. The method is safe, nearly 100% of the queens are mated. On mating stations the success rate is about 80% if the weather and food resources are fine.

Instrumental insemination makes the mating of the queens controllable, you know with certainty whom is the father. By this method you can make more crossings than in the mating stations where you only have one drone line. Only one genetic type delivers the paternal heritage, and there are no more mating possibilities than the number of mating stations.

Types of insemination
Sperm for insemination is taken from the drones and kept in a glass syringe. The sperm remains viable up to 35 weeks in a glass tube, which has been sterilized with streptomycine. It must be kept at 13-15^0 C. The sperm can be sent to other countries if you have the permissions. However, certain viruses can be transferred with the sperm.

Sperm cannot tolerate cold or freezing.

There are many ways of using the sperm. It is quite common to take the sperm from 10 or 20 drones living in the same hive and use it to inseminate a single queen. This makes the paternal heritage rather uniform. The disadvantage is that you get too few sex alleles which can heighten the mortality of the larvae unacceptably. (See chapter 13).

You overcome that problem by using sperm from many drones and homogenizing them by adding boar sperm diluter and centrifuge it. This mixes the sperm thoroughly. If you mix sperm from 300 drones you get enough to inseminate 30 or 40 queens. Every drone produces about 11 million sperm cells which are genetically uniform. In sperm from 300 drones there are only 300 different sperm cells which are mixed completely during centrifuging. The process is thus a homogenisation.

By insemination with homogenized sperm in a number of queens the paternal heritage is the same. The differences between the colonies of the queens must be caused by differences in the maternel heritage. This makes the choice of queens for breeding more safe.

Insemination makes maintenance of closed populations possible where the desired characters are kept. Surplus queens can be sold to beekeepers.

Insemination in genetic research
Insemination helps researchers in finding how certain characters are inherited. It can be done in several ways, for instance by inseminating the queen with sperm from a single drone or from two brother drones. You can even divide the sperm from one drone in portions and inseminate more queens with it. The queens will produce so many fertilized eggs that you can make a good judgement of the heritage.

Inbreeding is normally not used because bees are very susceptible to it. But it is done in scientific research. Here you can mate sister queens with drones from the same mother. It is also possible to inseminate a queen with sperm from her own drones: An unmated queen is anaesthesized twice with an interval of one day. After that, she lays unfertilized eggs and when they have grown up to mature drones their sperm is inseminated in their mother.

On techniques of insemination
Insemination is not difficult. But you need access to a laboratory where the drone sperm and the queens can be treated hygienically. The elaborated sterilization techniques makes insemination safe.

Insemination is learned on a course which lasts 3 or 4 days. The insemination of a queen, included sperm collection from 10 or 20 drones can be done in 10 minutes on the average.

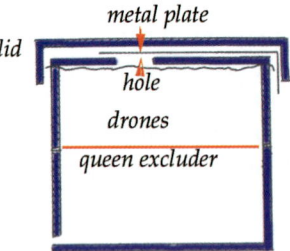

Fig. 47. Drones for instrumental insemination live in the super above the queen excluder. The lower storey contains a normal colony with a queen. The drones fly out through the hole at the top when the lid and the metal plate are taken away.

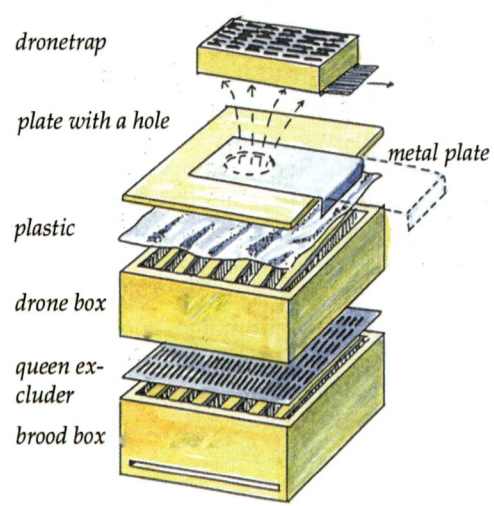

Fig. 48 Colony with drones for instrumental insemination. The lid has been taken away. The metal plate can be drawn out. Then the drones walk towards the light and get into the trap which has a lid of queen excluder and a movable metal plate as bottom.

Drones and their sperm

Drones are not so tied up with their colonies as the workers. Often the drones fly to foreign hives where they eat and stay. Because of that drones for insemination are not allowed to fly freely. In stead, full drone combs are placed in a super above a queen excluder together with combs full of pollen and honey (fig. 47). Much food is necessary for the maturation of sperm. The drones ought to be of approximately the same age.

The day before the insemination you can open the super of one hive in the evening after the normal flying time of the drones which is from 12 to 17 o'clock (summer time). The drones fly out, defecate, and return. After 15 or 30 minutes you can close the hive again. Even if you only have a small number of hives the work takes too much time because only one hive can be open at a time. Another solution is to catch the drones you will use and let them fly in a closed room. They fly to the windows where they defecate. Thus you are sure that the feces will not contaminate the sperm.

In the flying time of the drones you can collect them by using a trap (fig. 48). On other times of the day they are picked from the combs.

Insemination apparatuses

Insemination apparatuses have had a great influence on innovative people. There exist quite a lot, see the literature list. The illustration (fig. 49) shows the general principle. Most of the apparatuses have their supporters, in fact, if you have learnt the use of one of them you often will continue to use it.

The planning must be precise. The queens must be inseminated when they are 7-10 days old. The drones must be there and have plenty of sperm (1 μl per drone). The drones use 24 days from egg to adult. 12 or 16 days later it is ripe. So the planning of drone production must be as good as that of the queen.

Drone cells are constructed in April and May. An empty frame or a frame with drone foundation will often be filled with drone cells. The colony must be well fed.

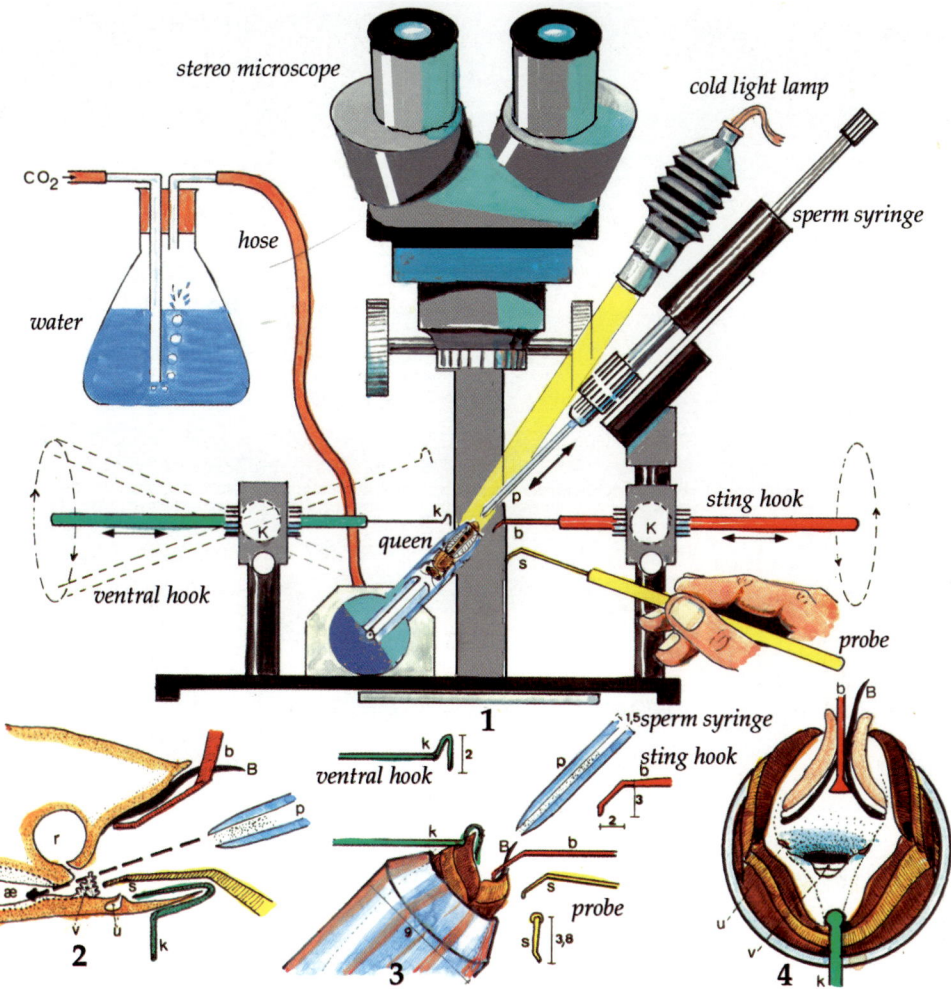

Fig. 49. **1.** *Insemination apparatus placed under a stereo-microscope. CO_2 from a container (not shown) passes the queen continuously and anestetisizes her. Before it arrives at the queen it passes through a bottle with water (upper left). Thus you can hear if the CO_2-stream is sufficiently slow. The CO_2 passes the queen and exits around her posterior. The tube with the queen can be turned until the queen is positioned towards the sperm syringe.*

A cold light lamp lightens up the tip of the abdomen.

The sperm syringe (p) first takes up sperm from drones (fig. 41, 55). The queen is fertilized when the sperm is forced out using the screw (upper right). The syringe can be moved forwards and backwards (double arrow).

*The ventral hook (k) is placed in the sting chamber and open it. After that the sting hook (b) pushes the sting (B) to the side (**2, 3, 4**). **1** shows that both hooks can be moved forwards and backwards (double arrow) and be turned in all direction in the sphaerical joint (K). The probe (s) is used to push the valvefold (v) in such a way that the syringe can go straight ahead (**2**).*

2. *The abdomen is opened. The spermatheca (r), the oviduct (æ) and the sting (B) are seen. The probe (s) is under its way across the opening of the bursa (u) and points towards the valvefold (v). After that, the probe presses the valvefold down (dashed line) and the syringe (p) is inserted in the oviduct (æ – fat arrow). The sperm is pressed out and the oviduct swells (dotted line).*

3. *The abdomen of the queen. Remark the probe (s) as seen frontally (right, lowest). Measures in millimetres.*

4. *The tip of the abdomen as seen through a microscope. The hooks are in place. The outer blue ring is the tube. B: Sting. The entrance to the oviduct is hidden by the valve fold (v). The entrance to the bursa (u) could erroneously be mistaken for the entrance to the oviduct. This is one example of an insemination apparatus; many other constructions exist.*

5. INSTRUMENTAL INSEMINATION 35

Fig. 51. Ruttner-Schneider-Fresnaye apparatus. Stereo microscope and cold light lamp above the apparatus. The parts are explained in the following figures

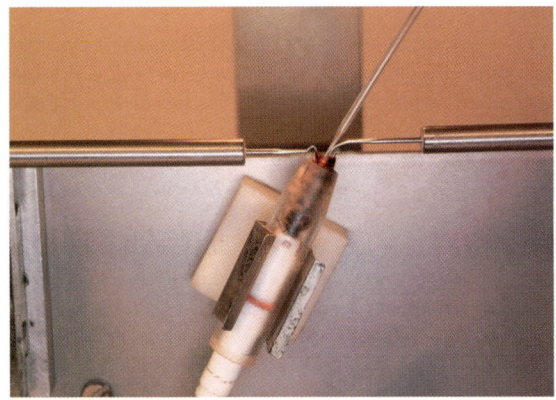

Fig. 52. Close up of the queen container. The queen is placed inside the plastic tube where she is anesthesized by CO_2 which gently is passing her. Her abdomen is kept open with the sting hook (left) and the ventral hook (right). The tip of the syringe is pushed into the oviduct.

Fig. 53. Penis. There is neither sperm nor mucus. This drone is sterile.

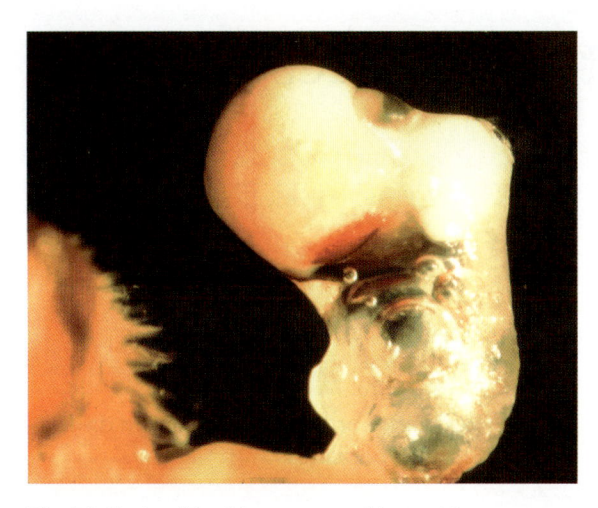

Fig. 54. Penis with white mucus and brownish sperm.

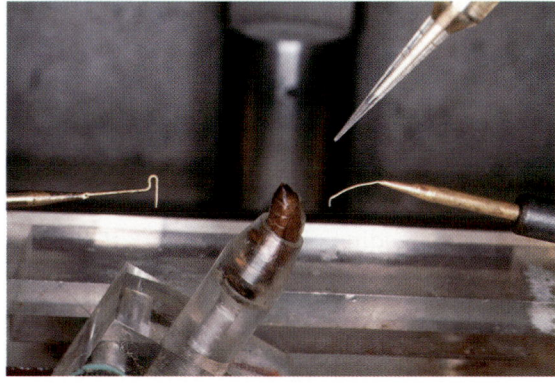

Fig. 55. Sperm is taken up in the syringe from the penis. You keep the drone between the fingers of the left hand while the right hand is turning the knob of the syringe. Compare fig. 41.

Fig. 56. The abdomen of the queen protrudes form the tube. It is surrounded by the ventral hook (left), the tip of the syringe, and the sting hook.

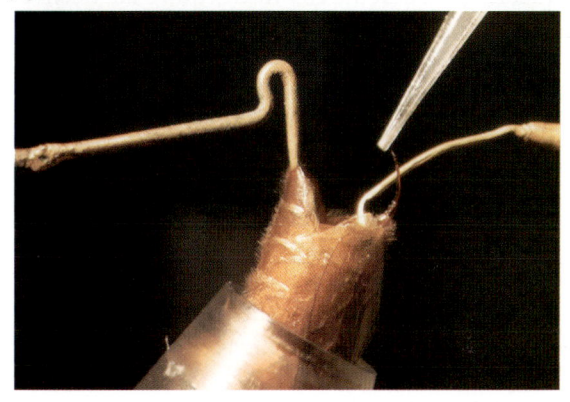

Fig. 57. The hooks keep the abdomen open. The tip of the sperm syringe points towards the opening of the abdomen. The curved sting is seen to the right.

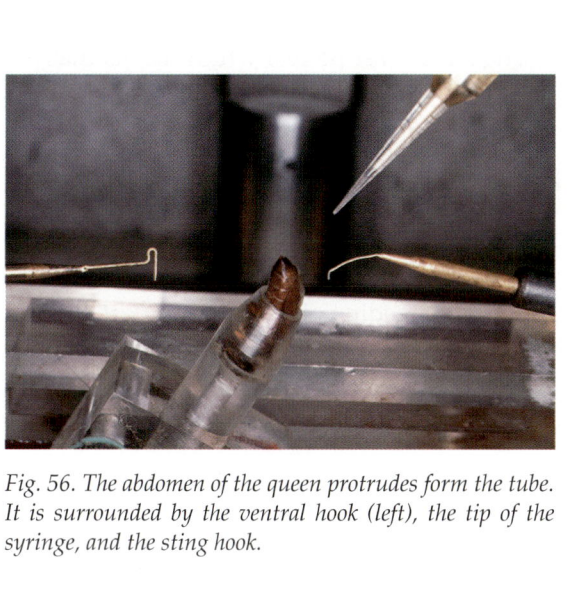

Fig. 58. The sperm is transferred to the opening of the oviduct.

Fig. 59. The queen is inseminated, the tip is inserted in the opening of the oviduct.

5. INSTRUMENTAL INSEMINATION   37

6. How to get better bees

Bee colonies are different, just like human families. Some colonies are gentle, other ones aggressive. There are vital and healthy colonies, while other ones are weak and sick. The honey production and the swarming tendency vary from one colony to the next.

Colonies can be good or bad from the point of view of the breeder. It is possible to breed the queens such that the colonies acquire the desired characters, more or less. The rest of the book describes how this is done.

A breeder of queens must be a good beekeeper. He or she must be able to judge himself by the honey harvest or the percentage of bees surviving the winter, the degree of infection by Varroa etc. The good beekeeper must give his bees the best environment inside the hives and outside. An important part of the work is the selection of queens. They can only show their good characters if their conditions of life is optimal.

The ABC of breeding
Breeding is done in steps. They are shortly described here to show the idea of the order of the chapters. Later on a full description is given in the single chapters.

The steps in breeding
1. The aims of breeding
2. Judging
3. Selection
4. Crossing
5. Testing
6. Maintaining

1. The aims of breeding
The breeder decides his own aims of breeding. The most usual ones are:
a. Honey production
b. Small tendency to swarm
c. Gentleness
d. Tranquility
e. Resistance or tolerance to certain diseases and parasites

Other aims are possible e.g. preference for certain flowers (white clover), winter hardiness, fast growth of the colony in spring.

The aims must be real which means that the desired character is found in different degrees in the colonies. The characters must be controlled by genes, at least partly.

The aims are described in the following chapter.

2. Judgment
When the breeding aims have been established you find the colonies which more or less have the desired characters. It is done by judgment of the colonies in at least one year. You note how gentle a colony is every time you examine it. The tendency to swarm is noticed, and so is the harvest of honey. The judgment must be objective, if possible. The number of swarm cells and the weight of honey are objective. You give a mark for each character during the examination of every colony. This makes it possible to choose the colonies to be used for breeding. See next chapter.

3. Selection
The selection of the colonies for breeding could be done by computers on basis of the marks. But the marks is only a part of the selection process. You ought to know the colonies personally by examining them through a long time. You must be careful in your choice because you only can use a small number of queens for breeding.

It is not enough to use the judgment of the colonies in the selection process. You have to use the pedigree of the colony, too. It shows if the desired characters are present and how they vary. Some characters can be present in a couple of generations, get lost in a generation, and reappear.

The pedigrees show which lines of bees or races which might be crossed with a good result.

Pedigrees are described in chapter 9.

4. Crossing

Crossing is used to make the offspring of your bees better. A much swarming colony with low honey production can possess a desired character, e.g. resistance to a disease. Then you cross drones from this colony with queens from your breeder colonies. The offspring will show a pattern of good and bad characters. Through selection and new crossings you might transfer the desired character to your breeder colonies without the undesired ones.

There exist many types of crossings described in the following chapters. Strategies of crossing is described in chapter 15.

The new lines you make by crossing and selection must be tested in different environments and be compared to existing lines. You need to be sure that the desired characters appear under different conditions. New lines should be better than the existing ones. If not, you reject them and begin a new.

6. Maintenance

During the breeding process you use insemination or mating in islands or isolated areas because unwanted matings must not occur. When you have got colonies with good characters which are present from one generation to the next they must be maintained. The matings are strictly controlled and carefully planned. The offspring is controlled, only the best are used in future breeding. However, the queens which are sorted out are often good and can be sold to the beekeepers. Thus you can finance the breeding.

The maintenance must continue so long time as bee breeding exists. If unwanted matings take place or the registration in pedigrees is not well done the good results of the breeding is lost in a couple of years.

Maintenance is described in chapter 15.

7. Aims of breeding and judgment

The queens you use for breeding are selected through judgment. The breeder has to find the answers to these two questions:

1. Which colonies shall yield the queens?
2. Which colonies shall deliver the drones?

These questions must be answered apart because we judge the ability of the queen by examining her daughters which are the workers and the queens. We test her worker offspring before we take new queens from the colony. We say: If the workers are good then their egg-laying sisters are good, too.

If we wish better information we must judge the daughter queens on their offspring before we decide to use their mother in breeding. But we can only compare the colonies of the daughters if they have been mated with drones from a number of sister queens from another stock, or if they are inseminated with homogenized sperm. In both cases the sperm is identical in the spermatheces of the queens. Now we claim that the maternal heritage is identical and so is the paternal heritage. Accordingly, the children have identical heritage. We know from our own children that this claim is not quite right. Neither is it quite wrong. If the mother and the father have identical characters (hair colour, body form etc.) there is a great chance that the children also get these characters.

The answer of the first question: "Which colonies shall yield the queens"? is: You take them from colonies with a 2 or 3 years old queen which has yielded good workers and good daughter queens.

The second question: "Which colonies should deliver the drones"? is answered thus: The drones come from unfertilized eggs. They only have the heritage from the queen, not from the drones which mated her. The drones of a colony have only the maternal heritage in common with the workers. But they have the same heritage as the workers in the colony from which their queen came. The answer of the second question is: If the workers of the colony from which the queen came are good the drones of that queen will be good, too.

The probability of a good result rises further if all the bees used for breeding have pedigrees three or four generations back. If all of the progenitors have been gentle and hard-working the offspring probably will have the same characters.

How to find good breeders
Breeding can only be done with success when many colonies can be compared to find the best queens. Therefore, the breeders must use the same basis of judgment and have the same aims. The customers of the breeders want to know and understand how the breeders are selected.

The common aims of the Danish breeders is to improve and maintain the following characters:

1. Low tendency to swarm
2. Gentleness
3. Tranquility
4. Honey production
5. Resistance to Nosema
6. Hygienic behaviour

Every character is judged by a scale from 1 to 5 (5 is the best).

Marks used for judging bee colonies

Made by Danish Beekeepers Association in collaboration with the Queen breeders' associations. Edition 2007.

Mark 5 is best, 1 is worst

1. Swarming tendency
5. No queen cells with eggs.
4. Queen cells with eggs. Nothing has been done to prevent swarming.
3. Queen cells with eggs. One preventive work has been done.
2. Queen cells with eggs and queen larvae which have been fed. More preventive works have been done.
1. Swarming.

2. Gentleness
5. Don't sting. Use of smoke unnecessary.
4. No stings when smoke is used.
3. 1 to 3 stings unprovoked.
2. 4 to 10 stings unprovoked.
1. Aggressive. Sting easily.

3. Tranquility (or temperament)
5. Very quiet. The bees walk seemingly undisturbed on the combs even if provoked. Can easily be shaken off. They don't take off even if shaken off.
4. Quiet. A little nervous movements on the comb. They don't take off, but they do fly if shaken off.
3. Nervous. The bees are running on the combs and a small number of bees take off unprovoked.
2. Inquiet. The bees run on the combs. Many bees take off unprovoked.
1. Very inquiet. Many bees in the air.

4. Honey harvest
5. More than 50% above the average of the bee yard.
4. 10 to 50% above the average of the bee yard.
3. The average of the bee yard +/- 10%.
2. 10 to 50% below the average.
1. Lower than 50% of the average.

Summing up when winter comes
1. Swarm tendency. Lowest mark of min. 3 judgments. Only mark 5-colonies are used in breeding.
2. Gentleness. Average of all judgments. Min. 5.
3. Tranquility (temperament). Average of all judgments. Min. 4.
4. Honey harvest. Only one mark is used.
   You must inform about the honey harvest of the colony together with average harvest of the bee yard and its number of colonies.

5. Nosema
5. No spores.
4. Very weak infection (0 to 0,5 million spores per bee).
3. Weak infection (0,5 to 2 million spore per bee).
2. Strong infection (2 to 5 million spores per bee).
1. Very strong infection (more than 5 million spores per bee).

6. Hygienic behavior
Freezing test (see text, no. 6, page 42)
5. Complete cleaning.
4. Nearly complete cleaning. More than 80% of the cells are cleaned.
3. Some cleaning. Between 20 and 80% of the cells are cleaned.
2. Weak cleaning. Less than 20% of the cells are cleaned.
1. No cleaning.

7. AIMS OF BREEDING AND JUDGMENT

Comments to the marks

1. Swarming tendency. When the bees of a colony swarm about half of the workers leave the hive after a period with little collection of honey. Some colonies swarm several times in the year and their honey production is low. The beekeeper uses much time to prevent swarming. However, swarming tendency is heritable. Danish bee keepers have reduced swarming by selection of colonies with low swarming tendency. It would be an error to breed bees which will not swarm at all, because swarming is the natural way of reproduction of the colonies. When beekeepers make new colonies they make an artificial swarm. If the swarming tendency disappeared completely the bees could not function as they should.

The aim is to make the swarming tendency so low that the beekeeper can control it by putting on more supers or making new colonies.

Swarming tendency should be controlled in all future. Good characters must be maintained, otherwise they are lost.

2. Gentleness. There is no reason to have aggressive bees. Gentleness is easy to get through breeding. In European bees, it can be done in a couple of generations. It is great not to have to use protective clothing. Heavy smoking is unnecessary but it is recommended to use a little smoke to "knock the door" before you inspect the hive. Even bees like politeness.

3. Tranquility. Working with the combs is easy if the bees remain on them during the inspection. It tells the beekeeper that bees are not disturbed by the work.

4. Honey production is compared with the average production in the bee yard in the same year. You cannot compare the production of different bee yards, or from year to year because the environmental conditions are too different.

The honey production is calculated by weighing the honey super before and after extraction. You could also weigh some combs and in this way learn to judge their load of honey. The production is described thus: 104/73 where 104 is the production in kilogrammes of the colony and 73 the average production of the bee yard. After that the mark is found according to the scale (page 41/4).

5. Nosema is a disease which sometimes causes heavy losses of bees or colonies. The resistance to Nosema is heritable, but cold and humid hives, or disturbances during winter can cause an outbreak. Danish breeders send 60 bees per colony, taken alive and killed in a freezer to the laboratory, "Sandagergård", where every probe is examined and the number of spores counted by a machine. The colonies get marks according to page 41/5. By using bees with the best results in breeding the disease has disappeared completely from many stocks.

6. Hygiene. The hygienic bees are described on page 64, 65, 66. The test for hygiene is made by taking a piece of comb with 100 sealed cells, freeze it for 24 hours, and replace it. 48 hours later you count the number of cells which have been opened and cleaned. The colony gets its mark according to page 41/6.

You do not need to replace the piece in the comb from where it is taken. You can take many pieces from the same comb and place every piece in a new colony. This makes the work easier. When you cut, you use a template which covers 100 cells. Before you cut you open a couple of cells to be sure that the pupae have not developed beyond the stage when they are still white with rose eyes. Be careful not to make a cut where there is a steel wire.

You can freeze the comb in the bee yard with frozen nitrogene. In stead of freezing a piece of comb you can use the *pin test*. You use the same shape as in the freezing test, but the piece of comb is only marked with colour at the top and bottom. Within the colour marks you kill 50 pupae with an insect needle, size 000, which is stung once vertically through the lid and the larva. 12 hours later you count how many cells which have been cleaned.

The hygienic behaviour is influenced by the weather and the possibilities of the bees to collect nectar and pollen etc. Because of that, you have to test the colonies which shall be compared at the same time.

Year _____	Hive no. _____
♀Marking _____	Bee yard _____
♀Pedigree _____	

			Judgments											
Date	Control of brood	Combs or supers	Buildup	Swarming (1-5)	Gentleness (1-5)	Tranquility (1-5)	Hygiene test (1-5)	Drone combs	Varroa treatment	Varroa count	Sugar food, kilogr.	Honey harvest, kilogr.	Probe for Nosema, date	Mark for Nosema	Remarks

Fig. 60. Score sheet for the Danish Beekeepers' Association. Buildup: Count the number of bee-filled spaces between the combs.

Marks of the year
After you have finished the work of the year you calculate the marks for every colony according to page 41.

How to use the marks
The marks is a judgment of the workers of the colony. From these we go backwards to the characters of the queens and the drones. They are thus judged indirectly, that is so close as we can come. But that method has given good results.

The marks follow the queens because her daughters are used as breeders. When you change the queen you cannot give marks to the colony until her workers dominate. Her first workers emerge from their cells 21 days after she laid her first eggs. Three weeks later they fly out to collect pollen and honey, and they continue to live a couple of weeks or some months (if they winter) before they die. The lifespan depends on the race and the time of the year. It takes many months before the new queen can be judged on her workers. If she came into the colony in August you must wait until June the following year until you can give marks to the colony.

You can easily see if a queen is so bad that she has to be changed. But she must be 2 years old before she is judged so thoroughly that you can determine that she can be used as a breeder. The greatest security is attained when her offspring queens have been judged and then she could be 3 years old.

The colony is judged by every inspection. The tendency of swarming only is judged in the swarming season. The Nosema test is made in April. The hygiene test should be repeated several times through the summer time. The production of honey is summed up once a year.

The Danish Beekeepers Association controls the quality of queens in the bee yards of the queen breeders. The queens are daughters of breeders and they are mated in a drone congregation area. Many queens for beekeepers are produced in this manner and testing of their quality is necessary. They get marks after the system on page 41.

Other judgments

Why have the breeders chosen the characters of the scale? Many other important characters exist, for instance early development, the time of maximum number of workers, lifespan of workers etc. The answer is that the characters of the scale can be objectively judged, or nearly so, and they are easy to examine. The choice of breeders is done by examining many series of sister queens and it should be made without too much work.

Every breeder of queens can examine the characters he wants so that he can choose his own queens. It is recommended that all interesting observations are noted because they could be used by the choice of the single queens. It might be a longer working day than the other colonies, eager collection of pollen, early building of drone cells etc.

When you inspect the bees you ought to use time to study them to learn the peculiarities of every colony. You get many impressions of behaviour, sounds, smell and sight which you cannot describe in words. In that way you obtain a set of experiences which is called *intuition*. It helps you to make the final choice of queens. *Note:* As a rule, you cannot explain your choice of wife or husband, or your friends; in these cases intuition is important, too.

The system of marks is very important and you can use it on a computer to help you to sort out the queens. When you have used the computer you are left with a number of queens or colonies from which you must choose. Here you use your own special notes and your intuition. Computers sort out, however, the final choice is your own.

Judges

You cannot judge without training. New judges are trained by judging together with experienced judges. The training must be maintained in a similar way.

The Danish Beekeepers Association uses consultants for the control of breeder queens. The same consultants judge the same bee yards during a long time in the presence of the owner. Thus the judgments become uniform, and the marks can be compared.

That control is deliberate and the breeders ask for it. Then they have to treat the colonies after a standard made by the association. Only 2 year old queens are judged. The breeder judges for himself during the first year.

Environment and judgment

Queens and their colonies which are going to be judged must have so similar environments as possible. It is very important when breeder queens should be chosen. The associations recommend:

1. The hives are placed so that the bees easily can find their own hives at the return flight (fig. 61).
2. Queen excluders should not be used because the colony should feel free to grow. There should always be a sufficient number of combs on which the bees can work.
3. You should not manipulate the colony through taking combs with brood away, moving such combs up or down in the hive, harvest honey in unusual times, use empty combs where the bees could build, use drone foundation etc.
4. Sister queens should live in bee yards with a different environment.

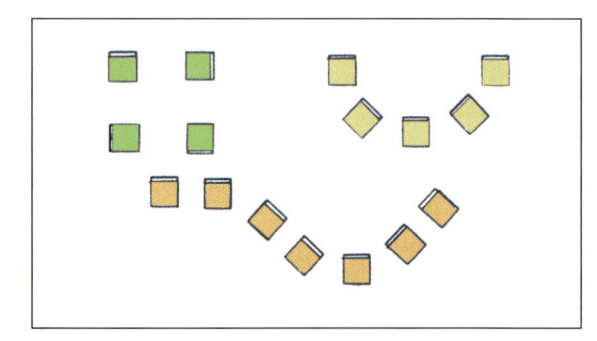

Fig. 61. Placing of hives to minimize the risk of bees flying to the wrong hive.

8. Queen Breeding and Genetics

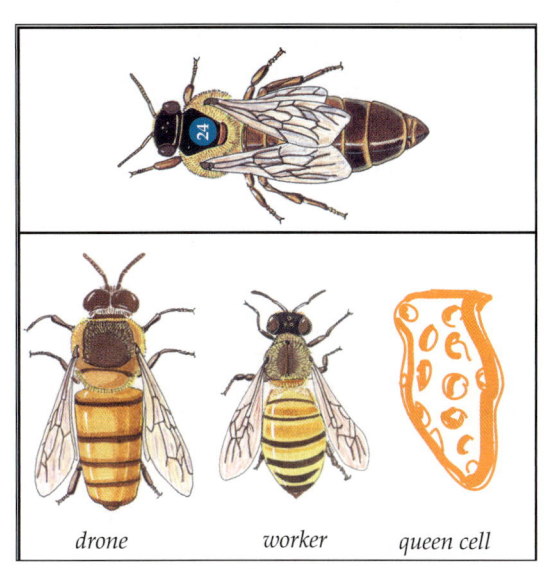

Fig. 62. The inhabitants of the hive.

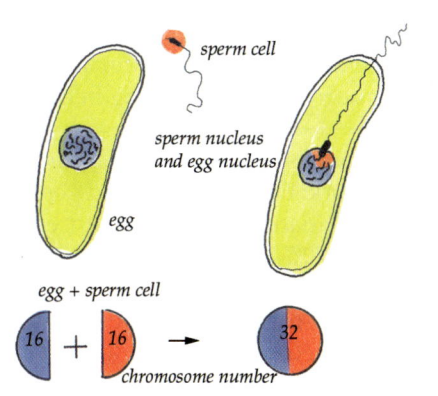

Fig. 63. Fertilization

Fig. 64. Diagram of a hive. Compare fig. 62. The somatic cells of the queens and workers are drawn as circles because they have 2n = 32 chromosomes. Egg, sperm and drones are drawn as half circles because they have n = 16 chromosomes.

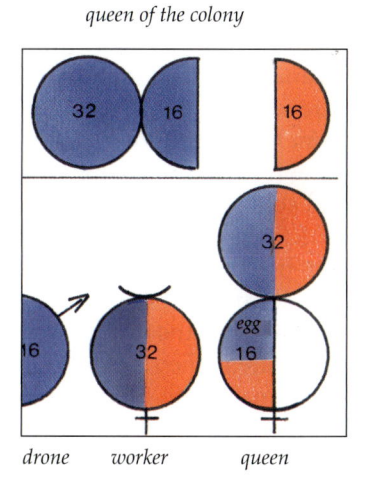

Honeybees follow the same genetic laws and rules as the other animals and plants, e.g. the Mendelian laws. The reproduction of honeybees must be understood if you want to use the universal laws of genetics in breeding.

A bee colony has a queen, workers and drones (fig. 62). Every bee has the same genome (= all of the genes on the chromosomes) in the unfertilized egg. If the egg is fertilized it becomes either a queen or a worker, depending on the environment (fig. 63). The queen egg is laid in a big cell and the larva gets special food, and because of that the larva grows to a queen. A worker egg is placed in a small cell and the nutrition of the larva is different from that of a queen larva. However, if you take a young worker larva and place it in an artificial queen cell it will grow up as a queen. The environment determines which genes should work in the larva.

The unfertilized egg has 16 chromosomes (n = 16, called the haploid number), while the fertilized egg has 32 chromosomes (2n = 32, called the diploid number). Fig. 63, 64.

The unfertilized egg is laid in a big cell and the larva becomes a drone. It mates with a queen and transfers sperm cells to her. Because of that, the drones are called male bees. *But it is wrong from a genetical point of view.* The explanation is, that the male sex cell is the unfertilized egg with its 16 chromosomes. This egg develops in a drone with 16 chromosomes in all of his cells. In his testicles he produces 11 million genetically identical sperm cells. The drone is just a copying apparatus, no new combinations of genes happens in him. It happened already when the queen made the egg cell. The new combination of genes is the important event in the formation of sex cells, it is explained further in chapter 12.

Who is the male, then? It is the queen. She is female when she lays a fertilized egg, and male if she lays an unfertilized egg. Accordingly, a

8. QUEEN BREEDING AND GENETICS 45

1   2   3   4   5 16

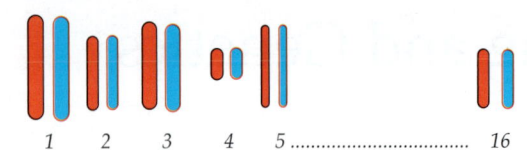

The 32 chromosomes in the somatic cells are paired, one in each pair comes from the mother, the other one from the father. When an egg cell is formed it must have 16 chromosomes, one from each pair. The selection can happen in such a way that the egg only receives maternel chromosomes (blue) or paternal ones (red), or a combination of paternal and maternel chromosomes.

Then are many possible combinations. Here four examples are shown in eggs from a queen. Her somatic cells and her egg-forming cells have chromosomes as shown in the upper figure. Only three chromosomes are drawn in each egg.

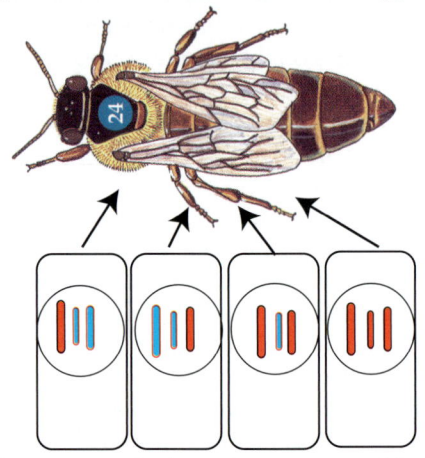

The two chromosomes of pair no. 1 are constructed in the same way. Genes for the same character are placed in the same place, or locus, on the chromosomes.

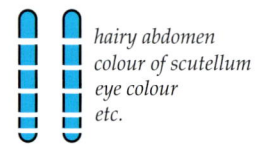

hairy abdomen
colour of scutellum
eye colour
etc.

The two genes can be different (alleles). Both influence the same character, for instance colour, but the colour of the bee can be brown, yellow or another colour.

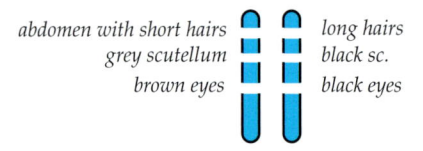

abdomen with short hairs   long hairs
grey scutellum   black sc.
brown eyes   black eyes

Fig. 65. The distribution of the chromosomes from a queen to her offspring. It is important which chromosomes the offspring receives. Because the chromosomes are selected in different ways everytime an egg is formed the sisters will be different to some degree.

queen is a hermaphrodite. The drones are her sperm cells, and the drones are as different as eggs or sperm cells in other animals. The differences between the individual sex cells are caused by the random distribution of chromosomes (fig. 65) and cross-over through meiosis (fig. 103), and it happened in the ovary of the queen. We could call the drones: *Sperm cells with wings*. Every drone makes millions of copies of the same chromosomes with their genes.

Bees, wasps, ants and bumblebees have the same way of reproduction: A diploid hermaphroditic queen and haploid males which make millions of identical sperm cells.

The advantage of this way of reproduction is that the drone is genetical similar to its sperm cells. When the drone grows up from egg through larva and pupa to adult a great number of its genes are tried out. If a gene is bad it will manifest itself at some time during the development of the drone, and it will never mate a queen. The life of the drone, then, is a genetic purifying process. However, it don't work for all of the genes, e.g. the genes for making of ovaries in queens, or the building behaviour of the workers.

In mammals another genetic purifying process is at work: Embryos with bad genes often die during pregnancy. The mortality can be quite high, up to 70% is reported in humans.

The genetic consequences of the mating behaviour
The honey bee is not tamed. It is a wild animal even when it lives in a hive, and it mates as it always has done in nature. In the mating season (from June to August in Denmark), the sexual mature drones fly away from their hives between 12 and 17 o'clock (summer time) if the temperature is above ca. 19^0 and the wind slow. Drones from many hives meet at a drone congregational area. It is said that drones from an area of 50 or 100 square kilometres gather there. If radius is 5,6 km the area is 100 km². A drone flies easily 30 km/hour = 1 km in 2 minutes.

The unmated queens leave their hives at the same time as the drones and fly to the congregational area. They fly lowly and the drones don't try to mate with them during the flight. Neither the queens, nor the drones know the place of the congregational area. However, they find it, and it is apparently the same through the years. It can be described as a place in the air 50 or 100 meters high and some hundred meters in diameter. We don't know how it is characterized or what the bees look for.

When the queen arrives at the congregational area, it flies upwards in the air, opens the sting chamber and releases a pheromone which attracts the drones. They come in their hundreds and form at "tail" behind the queen. Then the first drone mates, place its sperm inside the oviduct, its penis breaks, and the drone falls down, dying. Then the next drone arrives, takes away the penis of its predecessor, mates and dies. Then the next comes until number 5 or 7 or 10. Now the queen returns to the hive. She often alights on exactly the same spot on the hive from where she started her mating flight. Often a part of the penis from the last drone is seen on her posterior.

The next day she flies to the congregational area if the weather is good and mates again. Perhaps she flies out again a third time. However, if the weather is bad all the time she don't fly at all. If she does not mate she finally starts laying unfertilized eggs.

It is now possible to measure the number of matings of the queen. The method is genetic. It is the same as used by the courts to find out if a man can be or cannot be the father of a child. The method is described on page 55 (Microsatellites). In a research in France 196 colonies of bees were examined, and the number of drones with which the single queens had mated varied between 6 and 28. The average number was about 15.

If a queen has mated e.g. 15 drones then 15 different sperm cells are contained in her sperm container, the spermathece. There is room for about 5 million sperm cells there, but the queen gets much more. A drone contains about 11 million sperm cells. We don't know how she manages to take up a portion of the sperm of every drone with which she mates.

As all of the sperm cells from a certain drone are identical then all of the workers and queens from that drone have the same paternal genes, while the maternal genes are different from egg to egg. These workers and queens are closer relatives than e.g. human sisters and are called *supersisters*. Workers and queens with the same mother but with sperm from different drones are *half sisters*. We use this knowledge when we plan breeding of bees.

The hive has no members which are related in the same way as sisters or brothers in a human family. However, you can get them by inseminating a queen with sperm from drones from the same mother. Then you obtain supersisters and sisters. The drones will be just as different as the sperm cells of a man, while the eggs which they fertilize will be so different as eggs from a woman. Because of that the offspring is sisters when the sperm comes from different drones of the same queen.

The degree of kinship is important when inbreeding is planned. It is a method used in breeding, fig. 114. The degree of kinship is on the average 25% for half sisters, 50% for sisters, and 75% for supersisters.

Congregational areas

The congregational areas hinder inbreeding in bees. The chance of a queen being mated with drones from her own hive is minimal. A behavioural pattern which also fights inbreeding is that drones don't need to return to their own hive. They can go into foreign hives and are welcomed.

It is not easy to find a congregational area unless you are there while a queen is being mated. Then you can see the tail of several hundreds of drones behind her. But it only lasts a couple of minutes and between matings you only see some insects in the air without being able to identify them. On the Danish island Læsø some places with lots of drones were found by using a weather balloon with a catching net. It was provided with a black cigarette filter which had been soaked in queen pheromone. It attracts the drones and in some places many of them were caught in the net.

The short time of the mating process makes it improbable that insect eating birds will discover a congregational area and stay there to eat drones. The great number of the drones lowers the risk of the catching of queens.

9. How to make a pedigree

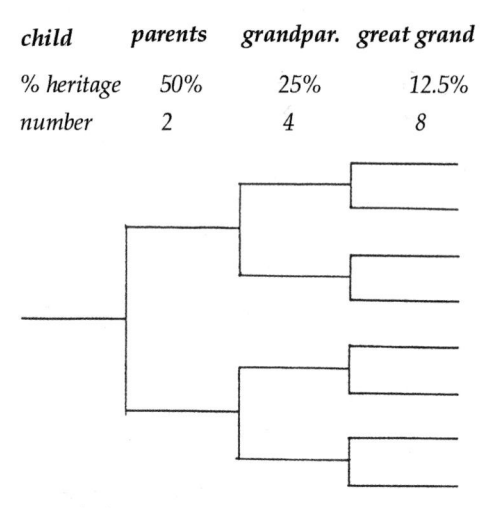

child	parents	grandpar.	great grand
% heritage	50%	25%	12.5%
number	2	4	8

Fig. 66. The human pedigree.

Breeding cannot be made without a pedigree. A pedigree has no meaning unless you know what the individuals in it have performed. It is also important to distinguish maternal and paternal heritage. The genes of the chromosomes are the same whether they come from the mother through the eggs or from the father through the sperm. However, the sperm nearly only contains chromosomes, while the egg contains both chromosomes, mitochondria with their own chromosomes and cytoplasm from the ovary with substances important for the development of the larva. This is important of many reasons which is described on page 56.

The human pedigree is very simple:
One person, two parents, four grandparents and so forth, as shown in fig. 66. The person has 50% of her mother's and 50% of her father's genes. That is a fact. Theoretically, the person has 25% of the genes of the grandparents, 12,5% of the great grandparents etc. Because the chromosomes from the parents are distributed by random during formation of the sex cells an egg or sperm cell can have e.g. 0%, or 10%, or 50%, or 75%, or 100% paternal chromosomes (fig. 65). Because of that we cannot say exactly how many chromosomes we have received from our grandfather or great grandfather. However, the heritage from our predecessors is rapidly thinned and we say that the heritage from the generations before the great grandfathers is of no interest in breeding.

The maternel and paternel chromosomes receive parts from each other during meiosis by cross-over (fig. 103). This complicates the way of heritage further.

Pedigrees show if two individuals are related and how much related they are. Then we can see if there is a chance of inbreeding.

Bees for breeding are judged according to their characters e.g. swarming tendency, peacefulness, disease resistance etc. If all of the predecessors have almost the same characters in the maternal and paternal line it is highly probable that the offspring will have the same characters.

A mating station is placed on an island or in an isolated area. The queens to be mated are sent to the station, their mother is known. The drones on the island come from 6 or more colonies living on the station, their queens are sisters (fig. 38). The cells of their ovaries produce the eggs which develop into drones, and these eggs of all the sister queens together represent the paternal heritage from their mother. It is paternal because it is transferred through the sperm of the drones from the sister queens. Because of that the mother of the sister queens is placed on the father's place in the pedigree (fig. 67). EH 431 and H 765 are the numbers of the queens.

┌─ her mother EH 431
│
queen to be mated on the island EH 314
│
└─ mated with drones from sister queens from HR 765

Fig. 67

The mother is placed on the upper line, her mating on the lower line.

The pedigree is written in this manner:

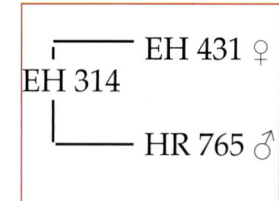

```
         ┌──── EH 431 ♀
EH 314 ──┤
         └──── HR 765 ♂
```

Fig. 68

The pedigree is continued: EH 314's offspring EH 412 is mated with drones from UG 507.

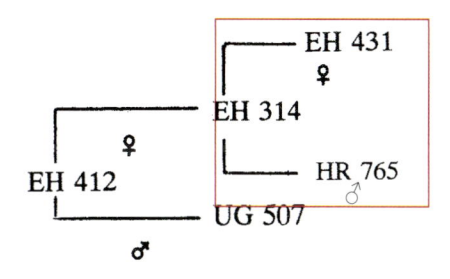

```
                    ┌──── EH 431 ♀
           ┌─ EH 314 ┤
           │        └──── HR 765 ♂
EH 412 ────┤
     ♀     └──── UG 507 ♂
```

Fig. 69

The pedigree is continued thus:

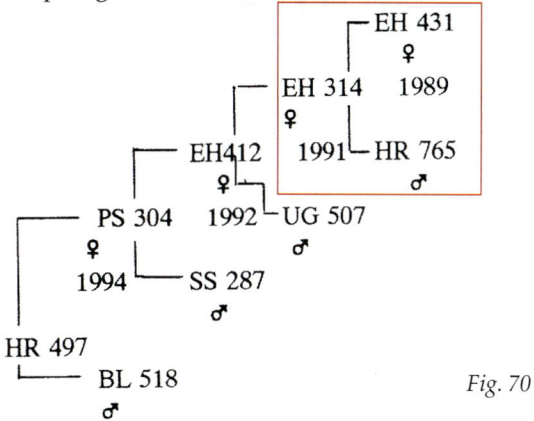

```
                              ┌─ EH 431 ♀
                   ┌─ EH 314 ─┤  1989
                   │    ♀    └─ HR 765 ♂
          ┌─ EH 412 ┤  1991
          │    ♀   └─ UG 507 ♂
  ┌─ PS 304 ┤  1992
  │    ♀   └─ SS 287 ♂
  │   1994
HR 497 ┤
       └─ BL 518 ♂
```

Fig. 70

The pedigree shows an unbroken line of queens and the mating which has brought the paternal heritage to every generation. The year of the matings is shown.

The Danish breeders pedigree

The Danish and Swedish breeders use a special way of writing the pedigrees. The method is originally invented by Brother Adam, the creator of the Buckfast Bee.

Here is an example, taken from the figure above:

HR497=.94-PS304xBL518:.92-EH412xSS287:.91-EH314xUG507: .89-EH431x765:etc

This is the pedigree of HR 497. HR is the initials of the breeder, 497 is the hive number which always has three ciphers. .94 is the year of the mating. Then comes the maternal heritage, always placed first (PS 304). After x its mating (BL 518). The shift of generations is marked by: (colon)
The combination of the breeders initials, the hive number and the mating year is the personal number of the queen; it can never be changed.

9. HOW TO MAKE A PEDIGREE

10. Bee cytology and genetics

Bees start life as an egg which is fertilized, except drone-eggs fig. 75). The cell divides many times until the adult bee is formed. It has millions of cells. The unfertilized egg has 16 chromosomes which contain the genes. The sperm cell, too, contains 16 chromosomes. The fertilized cell has 32 chromosomes. Every time a cell divides the chromosomes divide, too. The result is that every cell in the bee contains the same chromosomes and genes.

The cell divides by a process called mitosis, fig. 76.

The cells differentiate. There are nerve cells, muscle cells, gland cells and many other types. The differentiation is controlled by genes.

The cells have different functions although they have the same genes. Genes can be silenced in the cells where they shall not work. Other genes are waked up and produce e.g. wax in the wax glands when bees build cells. Here the demand of the bee colony triggers the genes of the wax glands. The time of the year triggers the physiological processes which make long lived

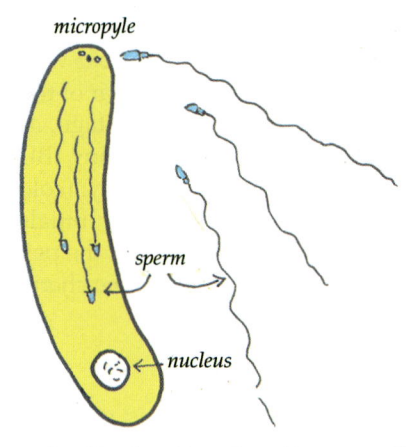

Fig. 75. Fertilization. From 3 to 10 sperm cells penetrate through the micropyle. The queen releases from 20 to 35 spermcells from her spermatheca per egg. Only one sperm cell fertilizes the egg and places its chromosomes inside the nucleus.

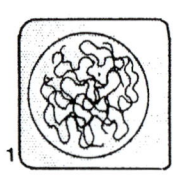

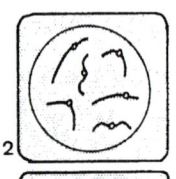

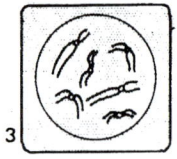

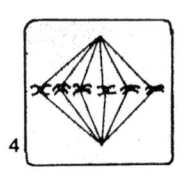

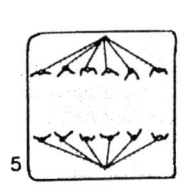

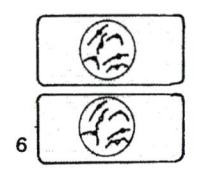

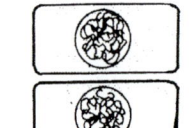

Fig. 76. Mitosis. Every chromosome divides in two after which the cell divides. The two cells receive identical copies of every chromosome. The chromosomes are easiest to study in the metaphase (4).

1. Interphase:
The chromosomes are invisible in a light microscope. They can be studied in the electron microscope. They are seen as thin strings. The interphase is the normal state of the cell where the genes work and make proteins. The division of the chromosomes is prepared in the interphase.

2. Prophase:
The chromosomes become easy to see and they can be stained.

3. Prophase (continued):
Now you can see that the chromosomes have divided. The two parts are held together by a centromere.

4. Metaphase:
The chromosomes have gone to the equatorial plane of the cell. The membrane of the nucleus is dissolved. A spindle of fibers grows from the two poles of the cell and bind to the centromere of every chromosome. It is easiest to distinguish the individual chromosomes in the metaphase.

5. Anaphase:
The cell begins division. The chromosomes migrate along the spindle fibres to the poles.

6. Telophase:
The nuclear membrane is retablished. The chromosomes become thinner and lose their ability to be stained. The cell divides.

7. Interphase:
The cells are back in the working state.

winterbees with much fat. Genes which control metabolic processes are often active many times a day.

The cell
A cell has a nucleus, cytoplasm and a cell membrane (fig. 77). Inside the nucleus are the chromosomes, which contain the genes. Pores in the membrane of the nucleus permit chemical substances to go from the nucleus to the cytoplasm and vice versa.

The cytoplasm contains much water with soluble substances e.g. amino acids, ions and sugars. Many chemical processes take place in the cytoplasm, catalysed by enzymes. The enzymes are made on demand. They are proteins and the formation of them are started and controlled by genes. The building of proteins take place in ribosomes, small bodies inside the cytoplasm.

The cytoplasm contains also mitochondria, there are hundreds or thousands of them inside the cells. Sugar, fat and other substances are broken down to CO_2 and water by the help of oxygen and enzymes in the mitochondria. The process frees energy from sugar or fat and transfers it to a substance (ATP). It moves to the places inside the cell where energy is needed and transfers the energy to the chemical process.

Mitochrondria contain special chromosomes. The genes on these chromosomes make enzymes for the mitochondria. The genes and enzymes of mitochondria are used to characterize bee races. (See page 80).

The genes
Inside the chromosomes are the genes (fig. 78). They are made of DNA. (DNA = Desoxyribo Nucleic Acid). There are also proteins inside the chromosomes which form a skeleton for DNA.

DNA is a very long molecule inside every chromosome. DNA is coiled inside a nucleus which is about 0,005 mm = 5 μm in diameter.

DNA looks like a spiraled staircase. It has two sides which are bound together with the steps. There are two coiled strands in DNA bound together by bases. There are 4 bases: Adenine, Cytosine, Guanine and Thymine, abbreviated A, C, G and T. Every "step" in the DNA-staircase is made of two bases, either A-T or C-G. We say that the DNA molecule forms a double helix.

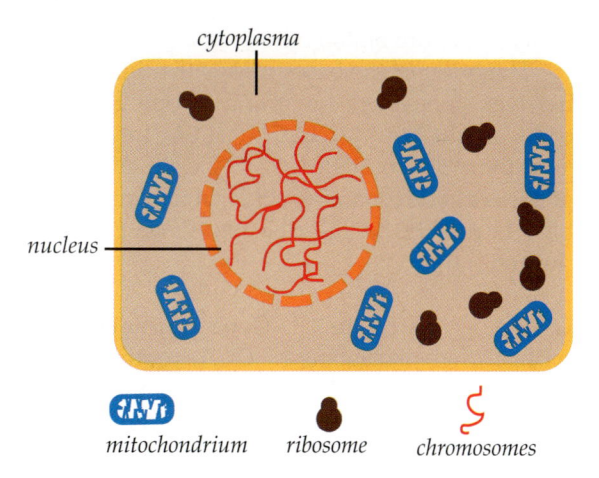

Fig. 77. A cell and its organelles. Only those described in the text are shown.

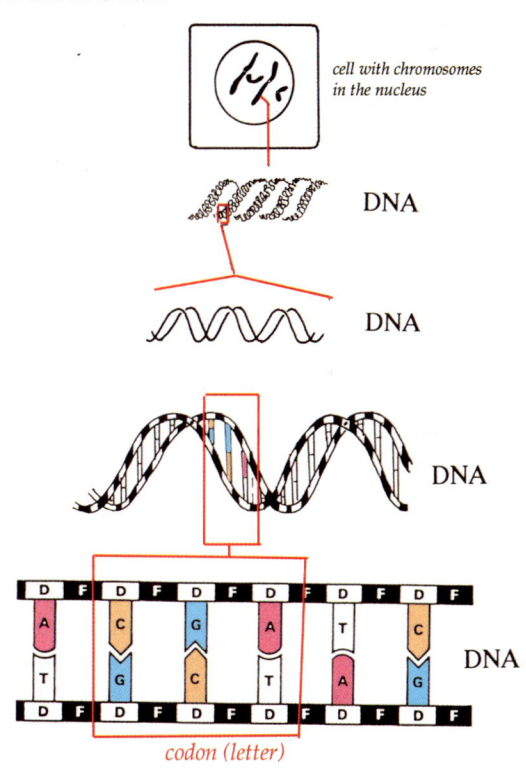

Fig. 78. Top: Cell with chromosomes inside the nucleus. Below: Part of a chromosome magnified more and more. Bottom: Part of the DNA molecule. A "letter" is framed.

Fig. 78 shows the base pairs. In one of the DNA-strands the bases are e.g. ACCTGA etc. The bases on the other strand form a complementary image of the first, because A always bind to T, and C always to G. The order of one strand determines the order of the other. If you only have one of the strands with its bases you can find out

10. BEE CYTOLOGY AND GENETICS 51

which bases are on the other.

The order of the bases makes the differences between genes. On the strand of one gene the order might be ATTCGAAGCTTA. In another gene the order could be CCACGACCGATG. The genes is a code for amino acids which are assembled to proteins in the ribosomes. We can read the genes thus:

Gene 1	protein 1	Gene 2	protein 2
ATT	isoleucine 1	CCA	proline 5
CGA	arginine 2	GTA	valine 6
AGC	serin 3	CCG	proline 7
TTA	leucine 4	ATG	methionine 8
...	...	...	...
...	...	...	...
...	...	...	...
...			

The numbered compounds are amino acids

The group of three bases is the building stone of the genes. Every three-group is called a *codon* (fig. 78) and codes for an amino acid. In all organisms from bacteria to insects and mammals 20 different amino acids are used to make the proteins. The genetic code has 64 codons. Some of the 20 amino acids are coded for by more than one base combination, e.g. proline is coded for by CCA and CCG. One of the codons codes for the beginning of a gene while 3 different codons code for the end of the gene.

The sequence of the bases in DNA are read by an automatic sequencer. Before that the material from the bee has gone through many procedures. The DNA has been extracted from the cells and certain parts of it has been amplified in a PCR-machine, and the DNA has been cut to pieces and marked with probes before the reading in the sequencer.

If a bee cell shall produce wax a chemical message is sent to the cells of the wax glands. The messenger is a chemical substance which goes through the cell to a gene starter (promoter) on the leading DNA-strand. Then a reading mechanism slides along the gene from the base-combination meaning *start* to the *stop* at the other end of the gene.

The reading mechanism makes a molecule which is a template of the leading strand of the gene (fig. 79). The template is called m-RNA

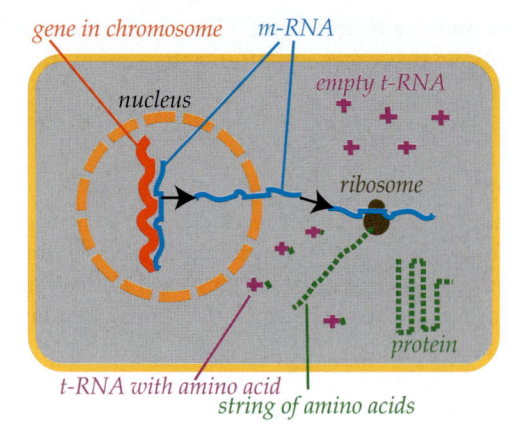

Fig. 79. From gene to protein. See text.

where *m* means *messenger*. It is not quite similar to DNA, because the enzymes of the cell must be able to distinguish between the genes and the messenger.

m-RNA goes through the pores of the nucleus to a ribosome in the cytoplasm. The ribosome is a small machine which slides along the m-RNA molecule, one codon at a time. As soon as it covers a codon a signal goes to the cytoplasm calling for a transport molecule which carries the specific amino acid to the ribosome. The transport molecule is called t-RNA and there is one for each type of codon. When a t-RNA arrives at the ribosome the amino acid is released and bound to the amino acid which came before. In this way all the amino acids are bound together in a long string. The order of amino acids is determined by the gene. The string is processed afterwards and a protein molecule is made.

There are many thousand different proteins in the cells of all organisms. All of them are made of the 20 amino acids. We can wonder that these 20 acids can produce so many proteins. However, the English alphabet has 24 letters and they can be combined to the millions of different words and sentences in English. So the alphabet of proteins are the 20 amino acids or the codons in DNA which control the position of amino acids in the proteins.

The work of the genes and the ribosomes is controlled. A chemical stimulus starts the production of m-RNA from the gene (fig. 80). Another signal stops the production. m-RNA is degraded when its production of amino acid chains must stop.

The anatomy of the gene

The gene is a part of the long double stranded DNA-helix. We can draw it simply in this way (fig. 81). The gene itself is combined with a promoter which receives the start and stop signals for reading of the gene. The start section tells the enzymes where to start the production of m-RNA. One or more enhancer sections can amplify the production of m-RNA. When much protein is fast needed at lot of m-RNA's are produced and many ribosomes produce the same protein.

Not all of the double stranded DNA helix is genes. Parts of it has apparently no function, it is called nonsense-DNA. We cannot explain why it is there but we can use it in many ways. Fig. 82 shows that it can be found between the genes and inside a gene splitting it up in coding and non-coding sections, called exons and introns. When the gene is copied to make m-RNA both introns and exons are copied. Special molecules cut out the intron m-RNA so that the m-RNA finally consists of the parts coded for by the exons.

We can distinguish the nonsense parts of a gene by looking at their bases. It can be ATATATAT….. AT tens or hundreds of times or CTCTCT……CT, or other combinations. There is much more nonsense parts in a DNA-double helix than genes. In the human DNA about 5% is coding, the rest is nonsense.

Fig. 81. A gene and its regions. See text.

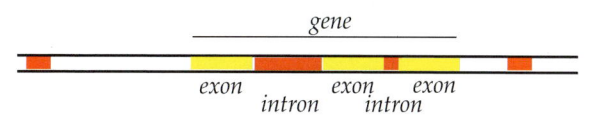

Fig. 82. Exons are the working parts of a gene. The introns (red) don't work and are cut away when the gene is copied to m-RNA.

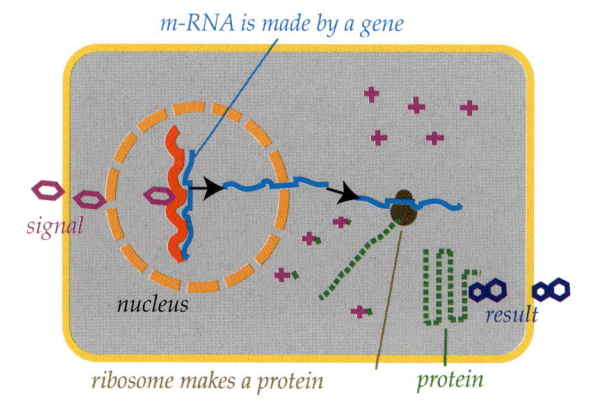

Fig. 80. A message from the outer world (purple polygone) makes a gene work. The result is af molecule (blue polygone) which is made inside the cell by the protein (enzyme) and exported. Compare fig. 79.

The genome and how to read it

The genome consists of the double-stranded DNA inside the chromosomes of the cell, both the coding parts which are the genes themselves, the promoters and enhancers, and the nonsense DNA. We talk about the human genome, the bee genome, the wheat genome etc.

The genomes are read by machines. They do not read a whole genome at a time. Instead the DNA is cut in small pieces by enzymes and placed in the machine which then reads the bases in order. Example: AGAGTCCCGGTA…… one reading. CCGGTAGAGCCCTAG…. another reading. There is an overlap (red) which is used to see which parts correspond to each other. Computer programs compare the readings and complete the design of genes and nonsense-DNA. Then comes the work of finding the chromosomes where the genes belong. It can be done in various ways e.g. by using a signal molecule which only will bind to one certain gene. The signal molecule is attached to another molecule with a fluorescent colour. When the chromosomes are placed together with the signal it binds to the gene. In the fluorescence microscope the signal molecule lights up with a special colour and you see on which chromosome pair it is found (fig. 83).

In 2006 the bee genome was published in a article by the *Honey Bee Genome Sequencing Consortium* in "Nature". According to that article most of the genes and the nonsense DNA of bees is known. You can read the codons in each gene, e.g. ACTTGACCACTG…. etc. You also recognize the codons of the nonsense DNA.

10. BEE CYTOLOGY AND GENETICS   53

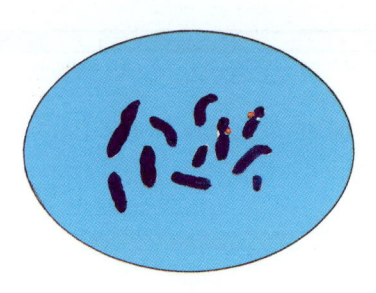

Fig. 83. Chromosomes stained with a chemical which binds to a certain gene. The chemical lights up when seen in a fluorescence microscope.

The next step is to find out what the genes do. When you know a gene you also can find the codons in m-RNA and the amino acid sequence in the protein made by that gene. So if you know a protein and its use in a cell you can go backwards and find the formula of the gene. Computer programmes do the work.

The function of many genes is found in another way. We know the genome of many bacteria, yeast, fruit fly *(Drosophila melanogaster, fig. 84)*, the malaria mosquito *(Anopheles gambiae)*, the chicken, the dog and man. More genomes have been described or are under way. By comparison, it has been shown that certain genes are common to most organisms, e.g. the genes for basal chemical processes are common to bacteria, plants, insects and mammals etc. All of the organisms have DNA and RNA. The chemical ways to make them and break them down is catalysed by nearly the same enzymes, which are made by nearly identical genes. The processes by which sugar is broken down or built up are the same, and so are the molecules (ATP, ADP) which transport energy inside the cell. By the help of computer programmes you can compare bee genes with genes from other organisms and find which of the known genes are more or less identic to those of the bee.

The fruit fly is the best known insect. It has been studied in the genetic laboratories for 100 years and an immense knowledge of its genes and genetic processes exists. By comparison with that insect many genes of the bee have been identified.

There exist databases, open for everybody, where the configurations of genes, m-RNA and the amino acid sequence of proteins can be found. They are updated currently and you can use them for comparison with your own results. A world wide collaboration is working.

According to *The Honey Bee Sequencing Consortium 2006* the comparison between the bee and other organisms has yielded many results. Quoting from the article: *The bee Apis mellifera has fewer genes for innate immunity, detoxification enzymes, cuticle forming proteins and gustatory (taste) receptors, more genes for odorant (smell) receptors, and novel genes for nectar and pollen utilization, consistent with its ecology and social differentiation…..*

…. Population genetics suggests a novel African origin for the species of Apis mellifera and insights into whether Africanized bees spread throughout the New World via hybridization or displacement.

In the years to come the function of many more genes will be understood. Most of the genes exist in several or many alleles (fig. 65, 85). The alleles are largely responsible for the differences between bee colonies, and in breeding we try to find the bees which have the most useful alleles. We can already do that with the breeding methods which are used today, we just look at the characters of the bees. However, we are not sure if these characters are genetic or are caused by some environmental factor. Many genes which influence the desired characters are not known, but the knowledge of the genome will help us to find them.

In genetic research a microarray is used. It is a glass plate with room for thousands of different genetic probes. DNA or RNA can be tested on a microarray where it binds to genetic probes. In this way you e.g. can compare genes which are active in bees producing wax, and which genes

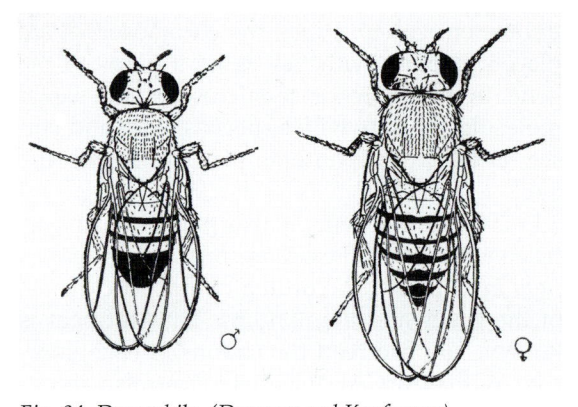

Fig. 84. Drosophila. (Demerec and Kaufmann).

are active in bees which do not produce wax. In this way you can find the gene which controls the production of wax. It is not known from other organisms. The microarray is read through the fluorescent microscope where the positive probes light up.

Microarrays for bee genetics exist in the laboratory. In the years to come the breeders hopefully will get a microarray to see the characters of their breeding material.

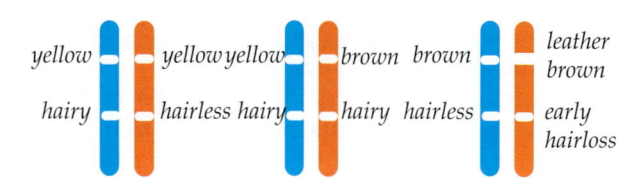

Fig. 85. Alleles are genes which exist in different forms. They are found in the same place (locus) of the chromosome and controls the same property.

Microsatellites

Microsatellites consist of a number of repetitions of bases in the DNA double helix. Examples: ACACACACAC.....AC (total 15 x AC). Another microsatellite: TGATGATGATGA......TGA (total 12 x TGA). There are many types of microsatellites, but they always have the same bases repeated many times. We don't know if they have any function, probably they have none. They are placed in different places on the DNA-double helix between or inside the genes.

Microsatellites vary in number of repeats. The example with 15 AC is found on a certain place (locus) on the DNA double helix. In other individuals there are 16, 17, 14 or 13 AC in the same locus. They are alleles (variants of the same gene).

In bees 474 different microsatellites are known. Everyone of them has more alleles. It means that microsatellite 1 (*m*1) normally has 15 repeats of AC (see table below). But there exist alleles with 16, 17, 14 or 13 repeats.

*m*2 has often 10 repeats of ATT, but alleles with 6, 7, 8, 9 and 11 repeats exist.

*m*3: 12 repeats of CGTA. Alleles 9, 10, 11, 13, 15 exist. Similar variation is seen in *m*4 and *m*5.

A sample of cells is taken from a worker bee and its DNA is extracted. This DNA goes through a number of procedures including amplification and the microsatellite is cut out using so called restriction enzymes. After that a probe molecule is tied to the microsatellite. Another probe is tied to another microsatellite. More microsatellites can be marked at the same time. The marked microsatellites are placed in an automatic sequencer which measures the length and thus the number of repetitions in more microsatellites.

Then the procedure is repeated for the next worker and you continue until you have examined the number you wish. See table below.

The counts for 4 workers starts with the maternel alleles. Because the queen has two chromosome sets, she also has two sets of microsatellites. Her workers get always one of the two maternel sets. The paternal set comes from the drone sperm. Workers 2 and 3 have identical paternal heritage what tells that they got it from the same drone. Worker 1 and 4 are different from 2 and 3. Here we see that the queen has sperm from at least 3 drones. If 50 or 100 workers are examined the number of drones with which the queen has mated can be found (from 6 to 28).

Mating Stations could be controlled in that way. You examine the microsatellites of drones from the station. The offspring from the queens which are mated on that station should only have paternal heritage from the drones of the island.

Number of bases in microsatellites in 4 workers

Microsatellite	from mother either or	worker 1 from drone	worker 2 from drone	worker 3 from drone	worker 4 from drone
m 1	15 14	15	15	15	14
m 2	10 10	11	13	13	12
m 3	7 9	10	9	9	10
m 4	17 19	17	18	18	16
m 5	12 14	15	15	15	14

10. BEE CYTOLOGY AND GENETICS 55

Maternel and parentel heritage

The eggs are produced in the ovaries of the queen. The nucleus contains the genes from her mother and father, and the same is the case for the nucleus of the sperm cells. The egg contains cytoplasm with mitochrondria, m-RNA etc. Neither m-RNA nor mitochondria are made of the genes of the egg. Instead, the m-RNA is produced by genes of the mother during the development of the egg and controls largely the development of the larva. Thus it is the genes of the mother, not those of the larva, which control the development in the beginning. It is called maternel heritage.

Another example is the mitochondria. There are hundreds of them in the egg. They divide independently of the chromosomes because they originally were symbiotic bacteria living inside their host cell. The egg gets all of its mitochondria from the mother and they divide when the cell divides. There are mitochrondria in sperm cells, however, they are rarely transferred to the egg.

The maternel heritage explains why it is important which bee-lines you use as mothers or as fathers. You cannot choose freely if you wants to use a certain line of bees as the source of queens or drones.

The properties of the queen influence the life and behaviour of her colony. If the queen only is able to lay, say, 1000 eggs a day her colony will get a life different from that of a queen which is able to lay 2000 eggs a day. The pheromone

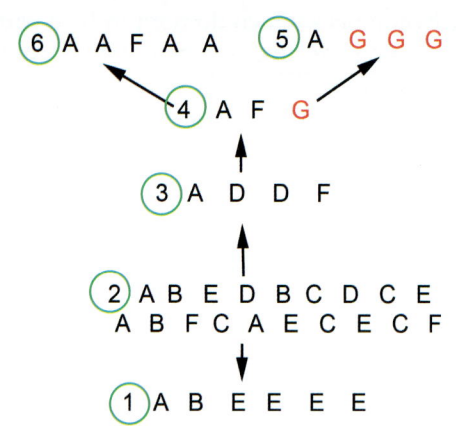

Fig. 87. Genetic drift. Every letter represents one genotype, for instance the offspring of sister queens. Every number represents a population. 2 is the original population, from which swarms have flown to the areas 1 and 3. Both have later been isolated from 2. Every new population has only a limited number of alleles with them. No. 4 has been created through emigration and isolation from 3. In 4 a mutation (G) happens. 5 and 6 emigrate from 4, but only no. 5 has the mutation.

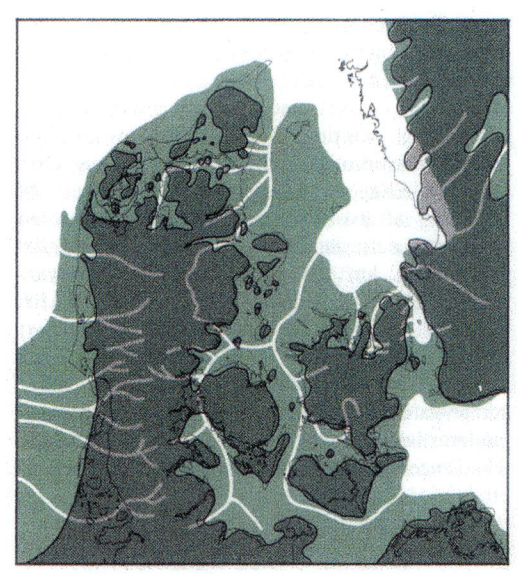

Fig. 88. Origin of isolated areas. The map shows Denmark about 9000 years ago when Denmark was a part of a continent together with Sweden, Northern Germany and England. Only rivers (white) separated the areas. About 9000 years ago the sea started to rise and it rose about 30 meters everywhere in circa 1500 years. Denmark was divided in islands (dark green) and the sounds became so broad that bees would not cross them. Later on the land rose and the northern islands were tied together. (Charlie Christensen 1993).

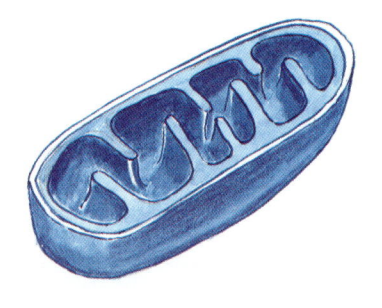

Fig. 86. Mitochondria cut through. It is less than 1/1000 mm long.

production of the queen influences the aggressivity of her colony. The more pheromone she produces the less aggressive are the bees.

Paternel heritage might consist in certain viruses which are transfered with sperm cells.

Mitochondria
Each of the mitochondria (fig. 86) contains 5 to 10 mitochondrial chromosomes. They are formed as rings and contain double stranded DNA. The genes of the mitochondria change very little or not at all from one generation to the next because mitochondria chromosomes are copied exactly when the mitochondria divides. As the mitochondria go unchanged from the mother to the next generation they tell the story of the population. There are changes by mutation but these happen rarely and by random. New forms will only dominate the population by a mechanism called genetic drift (fig. 87, 88). It works like that: If a small population of bees becomes isolated because of epidemic diseases or climatic changes, e.g. the Ice age, and if the mutations does not affect the function of the genes or happen in nonsense DNA, then, the bees with the change could survive by chance, and in this way start a new ecotype or race, while the "normal" bees die out.

Mitochondria are used in discriminating races, see p. 80, 81.

11. Meiosis and genes

Why are all queens, all workers and all drones different genetically? It will be described in this chapter.

The differences between individuals have their origin in *meiosis* and the *fertilization*. In the cells of the queen there are 32 chromosomes, or 16 pairs. 16 come from the mother, the other 16 from the father. The two of a pair are identical or homologue. Every pair of chromosomes has its own morphology and the same genes. Every pair is different from the other pairs. The chromosomes can be characterized by their form and content of genes. They are numbered from 1 to 16 (fig. 65).

By the formation of eggs in the ovaria the number of chromosomes is halved. The egg mother cell has 32 chromosomes and during two successive divisions of that cell the number of chromosomes is reduced to 16. This is meiosis. It is described in fig. 91.

By fertilization two sex cells unite and the original number of chromosomes is restored.

A queen has always two genes for every character, one comes form the mother, the other from the father. As earlier described the two genes can be different. They control the same character, e.g. colour, but the colour could be yellow, brown, grey, leather-brown etc. Such genes are called *alleles (fig. 85)*. An individual can only have two of the alleles, but many more exist. Alleles are the most important source of differences between bees and other organisms.

Dominant and recessive alleles
When queens from the brown bee of the Danish island Læsø are crossed with drones from the yellow Italian bee the drones of the offspring are brown but the queens and workers are yellow. This is caused by two alleles. The gene for yellow dominates the gene for brown. It means that the brown gene cannot be expressed when the yellow gene is there. The gene for brown is recessive, it can only be expressed if it is alone in the cells of the bee.

The drones have only one gene for every character because it is haploid (has only 16 chromosomes). If it has the yellow gene it is yellow, while the brown gene makes it brown.

Queens and workers have two genes for every character because all of their 32 chromosomes are paired. So they can have three different gene pairs for colour: yellow/yellow (yellow body), yellow/brown (yellow body) and brown/brown (brown body). The genes are written in this way YY, Yy, yy. Dominant genes are written in capital letters, recessive genes with lower case letters. The two genes in a pair are identified with the same letter (Yellow = Y, brown = y).

When the two genes are identical (YY or yy) they are called *homozygotic* (homo = identical). If they are different (Yy) they are called *heterozygotic* (hetero = different).

Fig 89 and 90. Yellow and brown (or black) bees.

1. Cell in interphase. The chromosomes are invisible.

2. The chromosomes condense and can be seen.

3. The chromosomes have divided. The two halves, the chromatids, are joined by a centromere. Every chromatid contains all of the genes.

4. The homologue chromosomes pair. One in a pair comes from the female, the other one from the male.

5. Two of the chromatids in each pair exchange parts by a process called crossover. This is the central event in meiosis because of its result: The four chromatides become different. This process is further described in fig. 103.

6. The crossover has finished. The exchange of parts between chromatides has been terminated.

7. The nuclear membrane dissolves and a spindle of fibres is formed from the poles of the cell. The fibres bind to the centromeres.

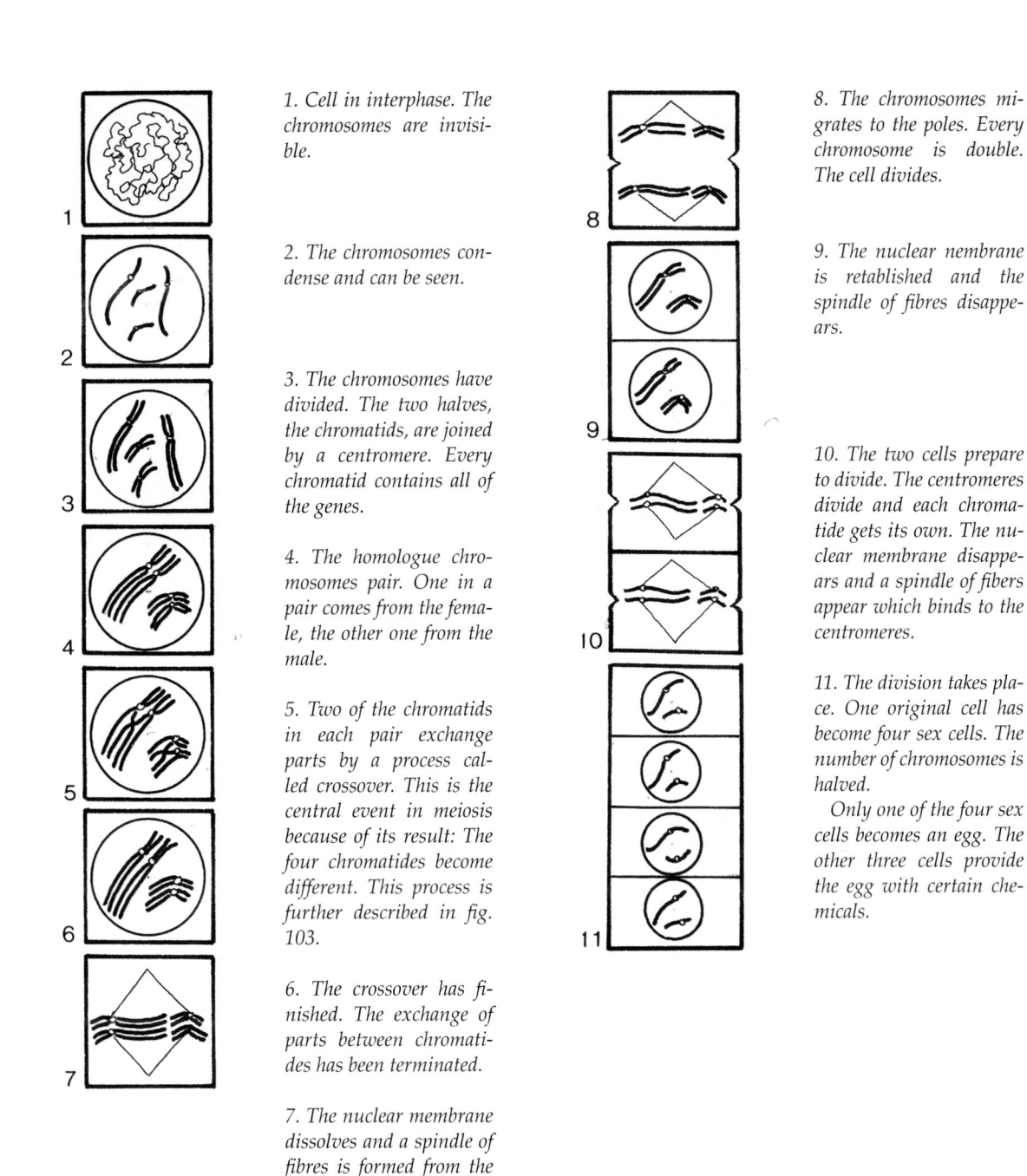

8. The chromosomes migrates to the poles. Every chromosome is double. The cell divides.

9. The nuclear nembrane is retablished and the spindle of fibres disappears.

10. The two cells prepare to divide. The centromeres divide and each chromatide gets its own. The nuclear membrane disappears and a spindle of fibers appear which binds to the centromeres.

11. The division takes place. One original cell has become four sex cells. The number of chromosomes is halved.

Only one of the four sex cells becomes an egg. The other three cells provide the egg with certain chemicals.

Fig. 91. Meiosis. An egg mother cell divides twice, the chromosome number is halved and the outcome is an egg cell.

Mating and segregation

On the island Læsø the brown (yy) and the yellow bee (YY) have mated. Later these matings have happened: Yy x yy and Yy x Yy. Fig. 94 shows the matings and their results. The capital letters below the hives are used in the following description. The hives are represented as diagrams with the queen above the horizontal line and its offspring below. The workers are omitted because we only shall see how the genes are transferred from one generation to the next.

Hive A and B are the keys to the following hives. The queen of hive A is a brown *mellifera* queen (yy) from the island Læsø. She has y sperm in her spermatheca. Her daughters are yy.

The queen of hive B is a yellow *ligustica* (YY) with Y sperm in her spermatheca. Her mating is without interest here because only her drones are used; they come form her unfertilized eggs.

The daughter from A (Yy) is inseminated with sperm from drones from B (Y). After that, the daughter is placed in hive C. Her drones are all brown (y) and her daughters are all yellow (Yy).

The eggs of the Yy queen are produced during meiosis. The result is 50% Y and 50% y eggs. The experiment is continued by mating two sister queens (Yy) from hive C. They are placed in hive D and E. D delivers queens for mating with drones from hive E. D is inseminated instrumentally with sperm from brown drones (y). The mating of the queen in hive E has no significance for the experiment as she only deliver drones. 50% of their sperm cells are Y, and 50% are y.

The unmated queens from D are either yellow (Yy) or brown (yy). Both of them are instrumentally inseminated with homogenized sperm from the drones of hive E (Y or y).

In hive F here are 4 possible combinations of genes:
sperm Y + egg y = YY (1)
sperm Y + egg y = Yy (2)
sperm y + egg Y = Yy (3)
sperm y + egg y = yy (4)

2 and 3 are the same because the result is the same whether the genes come from the mother or the father.

Crossing experiments of that type lie behind our knowledge of heritage. We work in this way:

1. As 50% of the drones belong to one type and 50% to the other one, then the heritage is caused by a single pair of genes, e.g. Yy (or Aa or Bb). See hive E.

2. Identical drones show that only one allele (Y or y) is present. Hive B.

3. When the young queens segregate in two types the cause is *either* that the queens are homozygotic recessive (yy) while the drones have two alleles (Y, y) – hive G, *or* that the queen is heterozygotic (Yy) while the drones are recessive, y.

4. When the queens segregate in three types in the ratio 1 : 2: 1 then the queen and the drones both carry the two alleles. Hive F.

This segregation is demonstrated in a scheme:

| Drone sperm | eggs of the queen |
	Y y
Y	YY Yy
y	Yy yy

Fig. 93 a

There are 1 dominant homozygote YY, 2 heterozygotes Yy and 1 recessive homozygote yy. Because of the dominance you cannot see which is YY or yy. The ratio accordingly is 3 yellow: 1 brown.

There is still an other important type which must be described: A yellow queen is inseminated with y-sperm from the drones of a brown queen (yy) - a test cross (fig. 93). If all of her drones are yellow (Y) and her young queens (and workers) all are yellow (Yy), then we know that she is YY, and she can be used as drone producer of Y-drones.

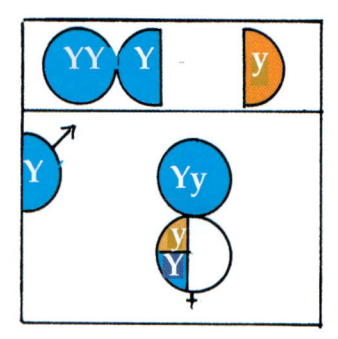

Fig. 93b. Diagramme of test cross. Compare fig. 94

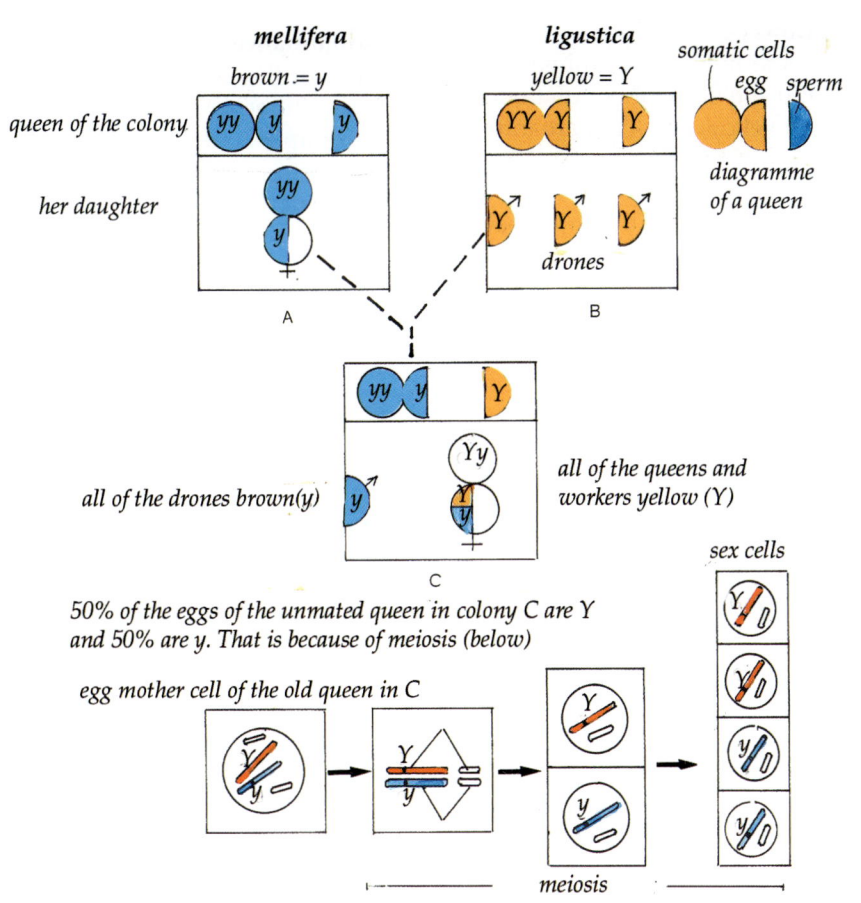

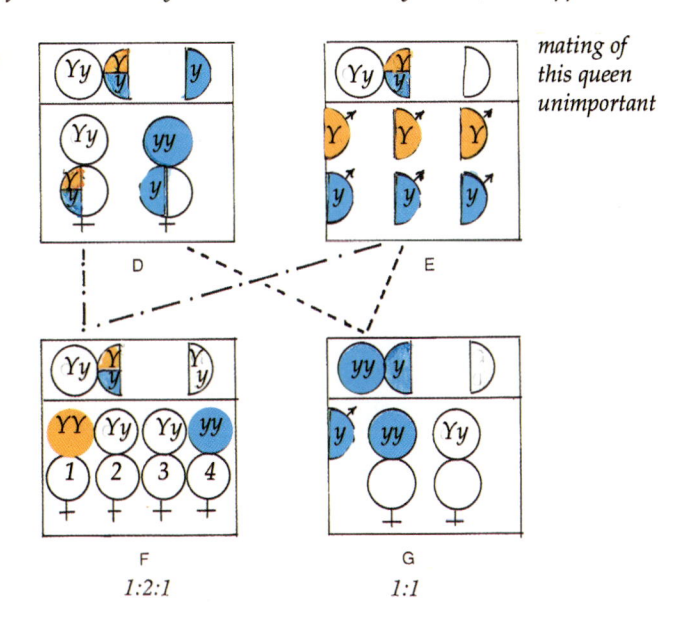

Fig. 94. See text.

11. MEIOSIS AND MENDEL   61

Mating in populations

All the bees which are living in a limited area form a population. If the original population is brown bees (yy) and yellow bees (YY) are introduced the proportion of brown to yellow bees depends on the number of the two types.

Example: In a limited area with brown bees so many yellow bees are introduced so that the proportion becomes 4 brown to 1 yellow. When the drones fly out to mate there are 4 brown drones per 1 yellow drone. The queens are brown. Figure 95 show the proportion of the colours in the offspring. It is 1 yellow to 4 brown.

Sperm	eggs of the queen Y
Y	YY
y	Yy
y	Yy
y	Yy
y	Yy

fig. 95

The following year yellow Yy queens fly out to mate. If a Gg queen mates with 20% yellow and 80% brown drones the result is as shown on fig. 96.

Sperm	eggs of the queen Y	y
Y	YY	Yy
y	Yy	yy
y	Yy	yy
y	Yy	yy
y	Yy	yy

fig. 96

The ratio of the genotypes is 1YY:5Yy:4yy while the ratio of the phenotypes is 6 yellow and 4 brown workers = 60%:40%. (The ratio of the drones of the offspring of Yy is still 1 yellow : 1 brown).

If the heritage of the alleles which are found in the workers of a bee colony is known you can find the ratio of the genes in the population by counting the worker types. You must examine many bee colonies to get a statistically significant proportion.

Co-dominance

If both of the genes in a pair of alleles are expressed in the organism there is no dominance. An example is the sex-alleles where the two in a pair both work to make a queen or a worker. This phenomenon is called codominance.

Mendel's first law

The genetic laws are valid for all organisms with sexual reproduction. The description of the crossing of the *ligustica* bee with the brown *mellifera* bee can be shown in this manner:

P-generation *(P = parental)*   YY x yy
Sex cells   Y   y

First filial generation (F_1)   Yy x Yy

Second filial generation (F_2)   1YY:2Yy:1yy
(filial = offspring)

You see that the genes which the offspring received from its parents are separated when F_1 makes its sex cells Y and y. Because they are separated they can combine freely. The result is that F_2 has 3 types while P only had 2 types. F_1 has only 1 type.

In practical beekeeping this knowledge is used in crossings. In F_1 every bee is identical when we look at a certain character if their parents were different, but homozygotic. The segregation in different types is seen in F_2 and only now you can select the queens which should be used for breeding.

Behind all this we find *Mendel's first law: The genes (Y and y) which the offspring received from its parents are separated when the offspring makes its sex cells such that 50% get Y and 50% y.*

Mendel was a Czech monk. He discovered this law and published it in 1865. He experimented with peas and other plants. However, the laws are also valid for animals.

Mendel worked with bees, too. However, his results were poor because the reproduction mechanisms of the bee were largely unknown. For instance, the mating of the queen with many drones was unknown, and so was meiosis and chromosomes.

12. Mendel's second law

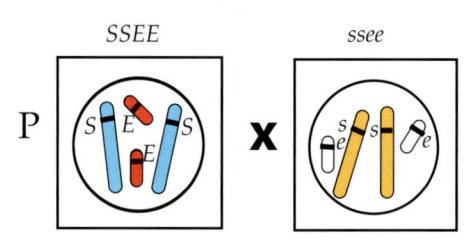

Segregation of two gene pairs

An attack of American Foulbrood (AFB) most often ends with the death of all of the bees in a hive because the larvae or pupae die inside their cells after the sealing. Some bee colonies are tolerant of AFB because they open the cells of the sick larvae and throw them out of the hive.

Two pairs of genes control that behaviour of the bees. There is a recessive gene for opening of the cells, it is called *s*. A dominant allele, *S*, works such that the bees don't open the cells. Another recessive gene controls the throwing out of the dead, it is called *e*. If the bees have the allele *E* they don't clean up the cells with the dead. There are four possible combinations of these characters:

1. The bees open the cells and throw out the dead. These bees are called hygienic.
2. The bees open the cells without cleaning them.
3. The bees don't open the cells, but if the beekeeper does it the bees clean the cells.
4. The bees don't open the cells, and if the beekeeper does it they do not clean the cells.

The ratio of the four types is seen if a hygienic bee (*ssee*) is crossed with a non-hygienic *SSEE* (fig. 97). The offspring (F_1) has the gene combination *SsEe* and point 4 behaviour. After that two F_1 are crossed: *SsEe* x *SsEe*. Every individual makes many sex cells. As shown on the figure the meiosis can run in two different ways (first and second possibility). 4 different sex cells are made in the ratio 1:1:1:1. They are seen in the scheme (fig. 98) which also shows the gene combinations. By sorting them out you find the ratio between the gene combinations of F_2.

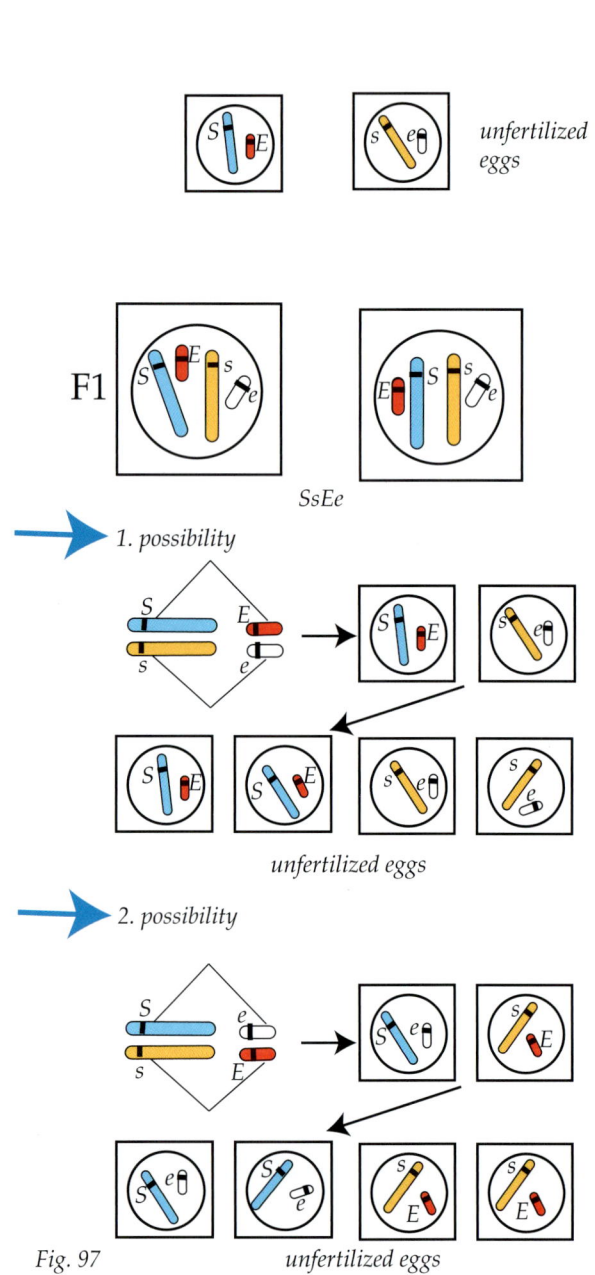

Fig. 97

drone	eggs of the queen			
	SE	Se	sE	se
SE	SSEE	SSEe	SsEE	SsEe
Se	SSEe	SSee	SsEe	Ssee
sE	SsEE	SsEe	ssEE	ssEe
se	SsEe	Ssee	ssEe	ssee

Fig. 98

Sorting

4 SsEe
2 SSEe
2 SsEE
1 SSEE
9 don't open, don't clean

1 SSee
2 Ssee
3 don't open, clean the cells

1 ssEE
2 ssEe
3 open, don't clean

1 ssee
1 open, clean

There are 9 different gene combinations, but only 4 types of behaviour. The 9 gene combinations are called *genotypes*. Because of the dominance you cannot distinguish between the different genotypes when you examine the behaviour of the bees. We say that the *phenotype* is the same for a number of genotypes. In this example there are 4 different phenotypes. The phenotype is simply what you can see. The genotype must be found by a test cross with the double recessive *ssee* or by using a microarray?

In this experiment you cannot look at the drones because they do not participate in the work with the cells and the larvae. You must look at the behaviour of the workers.

Only 1 of the 16 bees is *ssee* and is hygienic. You must use more than 16 queens to get the ratios of the scheme. Through testing for hygienic behaviour you find the most hygienic bees and hope that they are *ssee*. The result of the test is not alone due to the genes, environmental factors are working, too. What is needed are markers which show if a bee has the genes s and e.

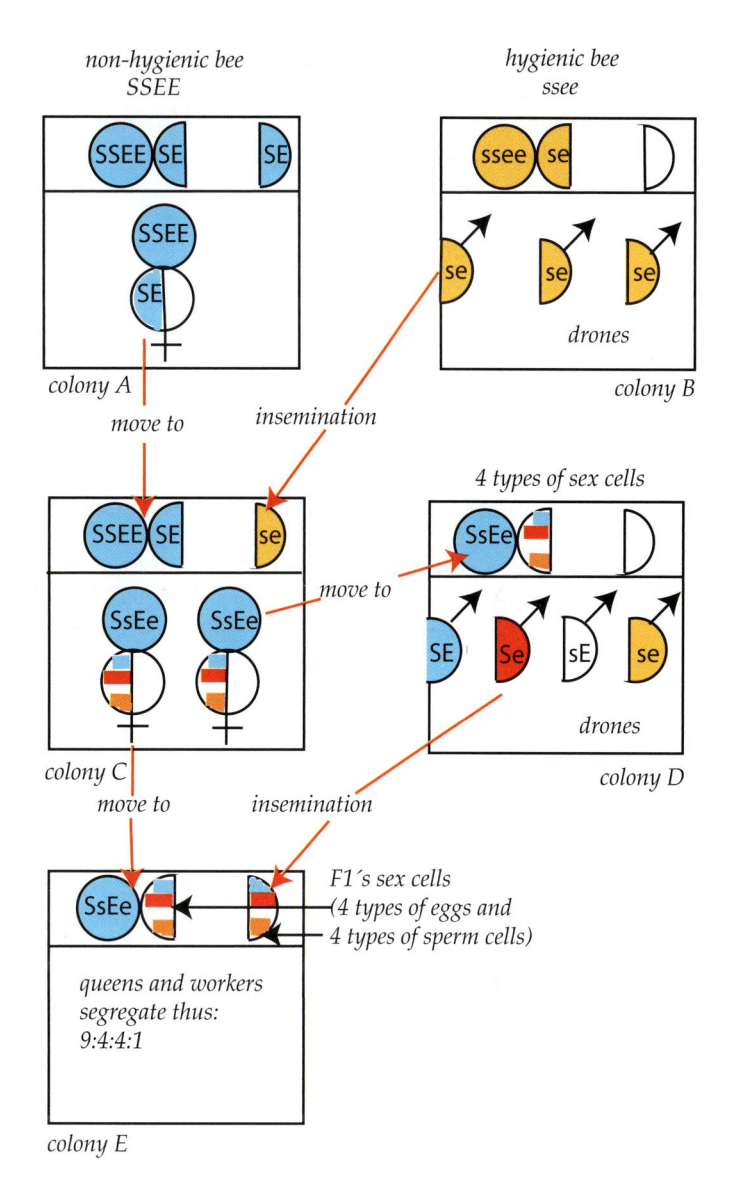

Fig. 99. How the crossing in fig. 97 and 98 is performed. Remark the use of insemination. Colony C and D have sister queens.

12. MENDEL'S SECOND LAW

When they are found sometimes in the future the work will be much easier.

You cannot find the hygienic bees by infecting the bees with American Foulbrood, it is too dangerous. Instead, you cut out a piece of the comb with sealed larvae and put it into a freezer where the larvae die. After that, you place the piece in that comb from where you took it. 24 or 48 hours later you examine the piece. You see that hygienic bees open and clean the cells. Other bees don't. If you open the cells they will be cleaned by bees which have that character but lack the ability to open the cells. This test is further described in chapter 7.

If you want to keep a population of hygienic bees you have to mate them with other hygienic bees on a mating station or by insemination.

If American Foulbrood infects a mixed population of hygienic and non-hygienic bees only the hygienic ones will survive. When AFB is fought you should not kill all of the colonies in a bee yard but let the disease free bees survive. After a couple of years the number of colonies have been restored and now all of them are hygienic unless non-hygienic bees have arrived through import, by migration, or free mating.

Another example: The original *ligustica* bee is resistant to the Tracheal Mite. It is probably due to a devastating attack in Italia after which only resistant bees survived.

Why are bees different
By comparison of the crossing schemes of Aa x Aa and AaBb x AaBb you can get an idea of how many geno- and phenotypes of bees can exist.

In the scheme Aa x Aa there are 1 pair of genes, 2 phenotypes (AA and Aa), 3 genotypes (AA, Aa, aa), and 4 places (fig. 93 a, fig. 100).

In the scheme AaBb x AaBb there are 2 pairs of genes, 4 phenotypes, 9 genotypes, and 16 places (fig. 97-99, 100).

If we continue the system with 3 and more gene pairs we get the results seen in fig. 100.

Example: The scheme shows how many different phenotypes which can be found if there are only one difference in each of the 16 chromosome pairs. In each pair of genes there is one dominant and one recessive allele.

The number of genotypes show how many different bees are seen if all of the genes are co-dominant. However, there are many hundreds or thousands of genes in every chromosome and because of that the possible number of genotypes is much greater.

Cross-over (see next paragraph) makes the variation still bigger.

Linkage and cross-over
The segregations of genes, described earlier, is true for genes which are placed on different chromosomes. If the genes are placed on the same chromosome they are linked (fig. 101) and the ratios of segregation changes.

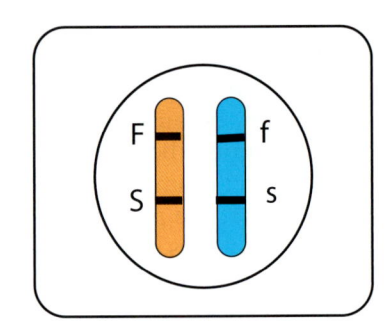

Fig. 101. Linked genes.

Differences in number of gene pairs	Phenotypes	Genotypes in the scheme	Number of places
1	$2^1 = 2$	$3^1 = 3$	$4^1 = 4$
2	$2^2 = 4$	$3^2 = 9$	$4^2 = 16$
3	$2^3 = 8$	$3^3 = 27$	$4^3 = 64$
4	$2^4 = 16$	$3^4 = 81$	$4^4 = 264$
16	$2^{16} = 65.536$	$3^{16} = 43.046.721$	$4^{16} = 4.294.967.296$
n	2^n	3^n	4^n

Fig. 100

The two genes *F* and *S* are placed on the same chromosome and are linked (fig. 101). *f* and *s* on the other chromosome pair are linked, too. *F* and *f*, *S* and *s* are alleles. In meiosis this could take place: (fig. 102).

The ratio is 1FS:1fs (50% of each). However, this is only an expectation. What really happens is that the chromosomes exchange parts by the process of cross-over (fig. 103).

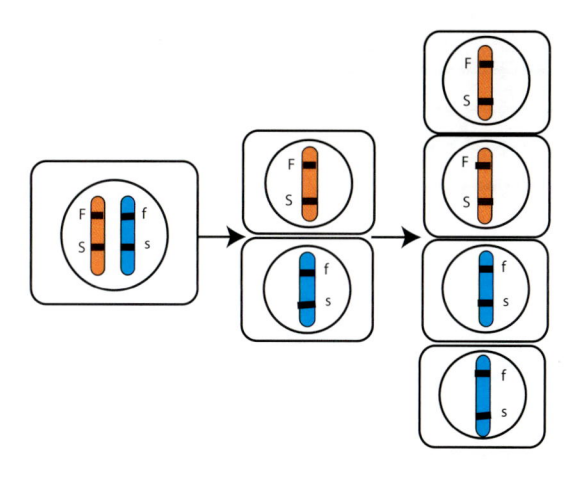

Fig. 102 A cell goes through meiosis and the result is 4 sex cells. Only two chromosomes with linked genes have been drawn.

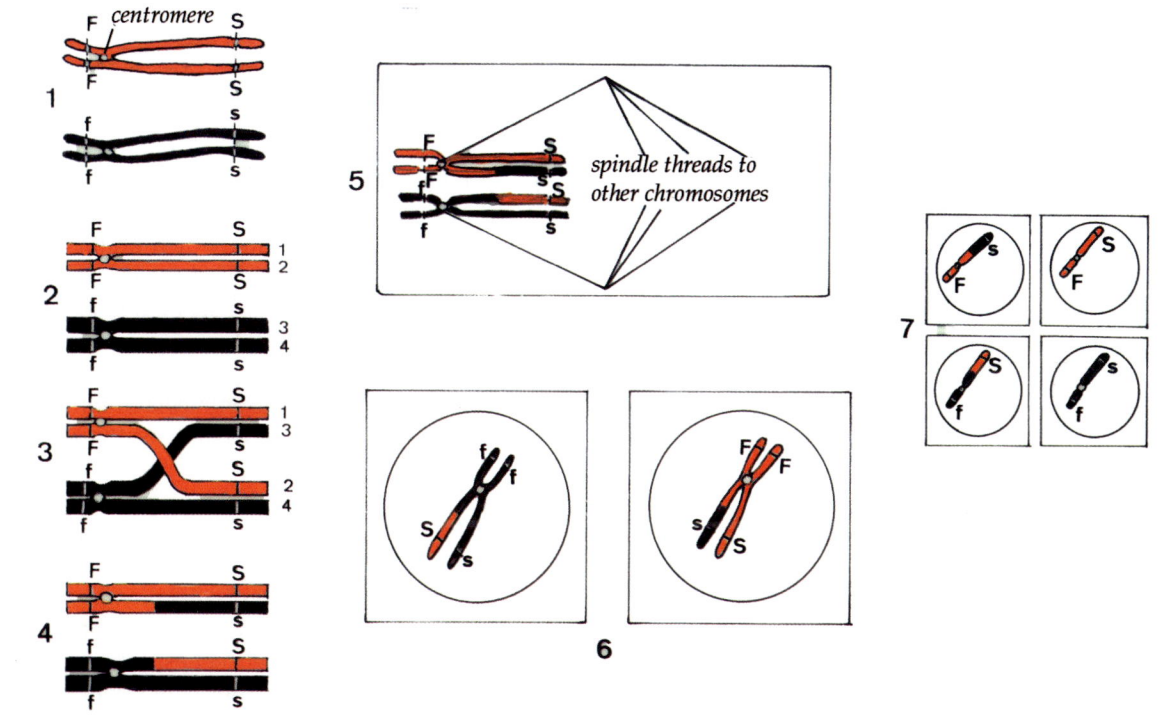

Fig. 103. Cross-over during meiosis. Confere fig. 91. Only one chromosome pair with two alleles is shown (Ff and Ss).
1. Cell before division (interphase). A pair of chromosomes is seen, each chromosome has divided preparing for cell division. See fig. 91,3.
2. The chromosomes have paired (see fig. 91,4). The division of the cell begins. The chromosome halves (chromatides) are numbered.
3. A short time after the pairing starts chromatids 2 and 3 make a cross-over. 1 and 4 don't cross-over. (See fig. 91,5).
4. During the cross-over the chromatids 2 and 3 exchange parts. After that the 4 chromatids are different.
5. After the pairing the chromosomes migrate to the equatorial plane of the cell and the spindle fibres are formed. (See fig. 91,6).
6. The chromosomes migrate to the poles of the cell which divides (fig. 91,9).
7. By the following cell division the chromatids are separated and become chromosomes. The sex cells are found. Because of the cross-over the sex cells are different. (Fig. 91,11).

12. MENDEL'S SECOND LAW

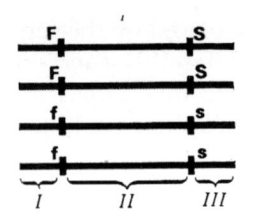

1. The same chromatids as in fig. 111,2. The chromatids are divided in three regions. Cross-over occurs in all of the three regions.

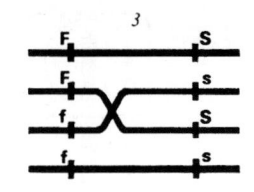

3. If the cross-over occurs in region II we get this result. The genes are replaced and we get four different sex cells: FS, Fs, fS and fs.

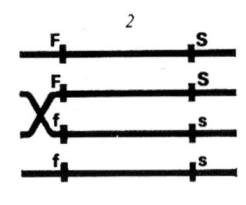

2. If the cross-over occurs in region I we get this result. The shown genes are not exchanged and only two types of sex cells are formed: FS and fs.

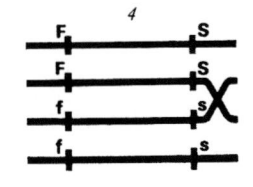

4. If the cross-over occurs in region III we only get two different sex cells, FS and fs.

Fig. 104. Different crossovers.

Cross-over is a normal process in meiosis but it occurs in different places on the chromosomes (fig. 104). The figure shows some possibilities. The closer the genes are placed the rarer is the cross-over between them. The ratio could be: 33,5% FS : 33,5% fs (no cross-over): 16,5% Fs : 16,5% fS (cross-over).

If the genes are closer to each other the cross-over % diminishes to e.g. 5% fS, 5% Fs.

Cross-over is very common in bees. On the average there are 4 cross-overs in every pair of chromosomes. It can be studied through the microscope.

Thanks to cross-over the linked genes will not be joined forever. Thus the number of gene-combinations is much higher than would be the case if cross-over did not exist. The importance of cross-over is so great that it must be a major cause if you seek an explanation of why meiosis is so complicated as it is.

If the bees were parthenogenetic the variation would be much smaller. The sexual reproduction combine the genes in new ways and is the cause of the many genotypes. There are so many possible combinations of genes that we are sure that every queen is different from all of the previous and future queens in the world.

What is the advantage of the great variation? Because of that every new generation of bees is different from the previous. The environment changes, e.g. the climate changes, new diseases and predators invade, and new ways of human use of the land develop. There will nearly always be new gene combinations which adapt to the new conditions and survive while the other die. This works only if there is a surplus of bee colonies in every generation which is the case when bees live in nature.

Every animal and plant with sexual reproduction vary just as much genetically as the bees.

Mendel's second law

The segregation of two gene-pairs obeys the second law of Mendel: *Different gene pairs in F1 (F and f is one gene pair, S and s the other) are distributed to the sex cells independent of each other.*

This is true for unlinked genes. Linked genes were discovered because they don't obey the second law of Mendel.

The law explains why segregation of two gene pairs gives 4 phenotypes and 9 genotypes in the F2 generation although the P-generation only had two types. If we examine the segregation of 3 or more genes the number of pheno- and genotypes fast gets immensely great, see fig. 100.

Mendel's second law is crucial in breeding.

13. Sex alleles, additive genes, nature and nurture

Sex is determined in many ways. In man sex is controlled by sex chromosomes: The male has one X- and one Y- chromosome while the female has two X-chromosomes.

Bees have no sex chromosomes. Instead, the sex is controlled by at least 12 sex alleles, but a bee will never have more than two of them. Alleles are placed on the same place or *locus* on a chromosome and control the same process. There are many examples of genes having two or more alleles, e.g. the gene for colour in bee races.

The 12 sex alleles function in that way: If there is one of the 12 sex alleles in the egg it develops into a drone (haploid). If there are two different sex alleles in the egg because of fertilization it develops into a female (diploid), either the fertile queen or the infertile worker. If the two sex alleles are the same in the egg it develops into a diploid drone. It can't function as adult. The larva produces a pheromone which urges the workers to eat it before it is 6 hours old. On its place there is an empty cell. If too many diploid drones are produced the bee colony becomes too .weak. The maximum loss can be so great as 50%. In experiments you can make diploid drones. You take a number of just hatched larvae in their cells, give them royal jelly, and place them in an incubator. 72 hours later you return them to the bees. Now diploid larvae are accepted, and they grow into diploid drones which are useless.

We need to take the sex alleles in consideration in breeding. A queen has two different sex alleles. Accordingly she will produce two different types of eggs. Inside her spermatheca is sperm from about 17 drones (average number). They could provide the eggs with other sex alleles. The congregational areas with their drones from many colonies makes it probable that the queen will get many different sex alleles.

In the following text and tables the sex alleles are numbered K1, K2 … K12.

When bees are instrumentally inseminated the drones can be taken from one hive. As the queen only has two sex alleles, 50% of the drones have one of the alleles, and 50% the other one. If the queen has K1K2, then the drones are either K1 or K2. The following scheme shows what happen if a K1K2 queen is inseminated with K1 and K2 sperm:

eggs of the queen	drone K1	drone K2	eaten larvae
either K1	K1 K1*	K1 K2	50%
or K2	K1 K2	K2 K2*	

* *diploid drone*
50% of the larva die.

Here are two other cases:

eggs of queen	drone K1	drone K2	eaten larvae
queen K2K3			
K2	K1 K2	K2 K2*	25%
K3	K1 K3	K2 K3	
queen K3K4			
K3	K2 K3	K2 K3	0%
K4	K1 K4	K2 K4	

* *diploid drone*

The number of eaten larva (seen as empty holes in fig. 108) is counted by using a transparent parallellogramme (fig. 107) which covers 100 sealed cells. If you count in May-June when the egg production is greatest you are sure of the result. The ideal is that there are no holes between the sealed cells, but it is rare. The following scheme shows what happens if a K1K2 queen is inseminated with sperm which contains from 2 to 12 sex alleles (fig. 109).

If all of the sex alleles in the sperm are found in equal proportions then the part of eaten larvae is 8,3% when there are 12 sex alleles. The scheme below shows how the number of sex alleles have been counted; the percentage of dead larvae falls when the number of sex alleles gets bigger. There might be more than 12 sex alleles, but the method cannot show that. The difference between 12 and 13 sex alleles is so small that it cannot be seen with certainty.

Effect of the number of sex alleles

There are several ways of getting around the problem of mortality because of too few sex alleles. When you use instrumental insemination you use drones which are not related to the queen. You should not take drones from one hive but from a number of hives. Some inseminators mix sperm from many drones from different hives. The mixing is done in a centrifuge.

When you use sister queens of drone producers, then half of them has one of their mother's two sex alleles, while the other half has the other. If the mother queen has many different sex alleles in her spermathece the other sex allele in the sister queens will probably be different. If you measure the percentage of eaten larvae in the hive of the mother queen you see if she has many or few sex alleles (fig.107). If she has too few she will not be used in the breeding.

Inbreeding in bees result in a reduction of the number of sex alleles. How this problem is solved is described in chapter 15.

queen K1K2

her eggs K1 or K2

drone sperm with	*drone sperm with*
3 sex alleles	*4 sex alleles*
K1K2K3	*K1K2K3K4*

	K1	K2		K1	K2
K1	*K1K1	K1K2	K1	*K1K1	K1K2
K2	K1K2	*K2K2	K2	K1K2	*K2K2
K3	K1K3	K2K3	K3	K1K3	K2K3
			K4	K1K4	K2K4

33% eaten larvae | *25% eaten larvae*

If you continue in the same way the % eaten larvae diminishes:

number of sex alleles	*% eaten larvae*
5 K1-K5	20%
6 K1-K6	16,67 %
7 K1-K7	14%
8 K1-K8	12%
9 K1-K9	11 %
10 K1-K10	10%
11 K1-K12	9%
12 K1-K12	8,3%

Fig. 109. The number of eaten larvae depends on the number of different sex alleles in the sperm.

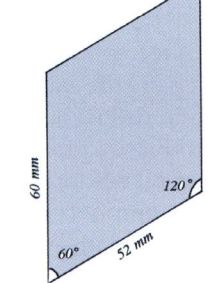

Fig. 107. Parallellogram of transparent plastic covering 100 worker cells.

Fig. 108. Failure of egg hatching is seen as holes among sealed brood. The failure on this comb is not alarming.

None of these methods are exact. It should be possible to develop a microarray which can measure the number of different sex alleles in a bee colony using a probe of DNA from, say, 100 workers.

Additive genes
If a character is controlled by one (*Aa*) or two pairs (*AaBb*) of genes you can see it on the number of phenotypes in the offspring after a crossing. It can also be found when you count the number of phenotypes = genotypes in the drones of a family.

Yield of honey, length of life, certain types of disease resistance are controlled by a great number of gene pairs, perhaps 5 or 10 or more. An example shows what happens in such cases. It is taken from wheat because no clear examples exist in bees.

In a certain type of wheat the colour of the grain varies from dark red to white. There are a number of shades of red in between. The deep red has the genotype $R_1R_1R_2R_2R_3R_3$. R is a gene for colour. In the chromosomes of wheat there are 3 pairs of genes which all are taking part in the colour. That type of genes are called *additive*. The white colour in grains has the genotype, $r_1r_1r_2r_2r_3r_3$. *r* is a gene for the character *colourless*.

A plant with 5 R-genes and 1 r-gene has red grains, not so dark red as that with 6 R-genes. If there are 4 R-genes and 2 r-genes the colour is lighter red etc. It is remarkable that plants with $R_1R_1R_2R_2r_3r_3$ have the same colour as $R_1R_1r_2r_2R_3R_3$ or $R_1r_1R_2r_2R_3R_3$. What really counts is the number of R-genes.

It was possible to analyse what happened in wheat because there only are 7 types of colour. If the colour had been controlled by more pairs of genes the colour groups could not be distinguished with certainty. Only a scale from white to red could be made. It is a characteristic of additive genes that the characters they control go from one end to the other on a scale. Yield of honey varies from few to several hundred kilograms, and you cannot divide the bee colonies in groups, which could tell us the number of genes involved. Disease resistance varies between: *All of the bees have died* until *All of the bees survived*.

The characters, which are controlled by additive genes, are important and we need to know if a character is controlled in that way. Keeping many different bee colonies in a uniform environment will show it. If the character varies smoothly additive genes control it.

Every pair of genes is inherited in the usual way. We have no methods which show the number of involved genes. Perhaps the study of the genome will help.

Aggressivity in bees might be controlled by additive genes, at least in some cases. The breeding for gentelness shows it. If you take the most gentle bees and let their queens mate with drones from gentle lines of bees you get more gentle bees in each generation. Sometimes selection for gentleness works so fast that the beekeeper don't need protective clothing after three or four generations.

You could make aggressivity greater by selection of the most aggressive bees and end up with "killer bees". Selection of characters controlled by additive genes work in that manner. Sometimes there is still progress during selection even in 50 or 100 generations.

In some cases additive genes probably cause resistance to diseases.

Mutations
Mutations are changes of the genes. If only one base pair changes an amino acid in the protein can be exchanged with another, and the protein can change or lose its function. Alternatively, if the result of the mutation is a stop codon, then the transcription of the gene is stopped where the "stop" is. The protein from that gene cannot work if an essential part is missing.

Copying errors. When a cell divides the genes are copied. The copy of the two strands must be without errors. Several mechanisms control the copies and compare them with the strands from which they are copied. If an error has happened it is corrected. Certain control molecules slide along the copy looking for holes in the row of codons and closing them. The controllers can also find a wrong base and put the right one in its place. Although these mechanisms work with extreme precision errors can happen. Then a mutation has been created.

Sometimes a gene is duplicated during the division of cells. If a certain gene is necessary

Fig. 110. Queen and her offspring. She has the mutations cordovan, diminutive and umber which give light body colour, short wings and amber coloured eyes. Her offspring are different according to which father they have. They are either normal or have all of the three mutations or one or two. (Per Kryger fot.).

for the life of the organism a mutation might be catastrophic. When two identical genes exist in the same genome then one of them could mutate and make new proteins while the other keeps the original function. There exist genes with related genetic codes making similar products. These gene "colonies" might have evolved in this way.

Chromosomes can break and anneal in new ways. Even if no genes are lost this process can change the performance of them. Its company of other genes, promoters and enhancers influences the function of a gene.

Mutations are the common name of all these genetic changes. They often happen when cells prepare to divide.

Every gene in every cell can be touched by mutations. If they happen in the egg producing cells they might be carried on to their offspring and further through the generations. If mutations happen in all of the other cells (the somatic cells) they can damage them. In mammals a mutated cell can develop into cancer.

Mutations in the egg producing cells are often deleterious and the offspring could die or be sick. Some mutations are of no concern e.g. when a codon is changed to another one coding for the same amino acid. Some mutations are good. The proof is that all alleles are made by mutations.

The genes have different mutation rates, generally they are between 1 in 1 million to 1 in 10 billions.

Mutations are caused by many factors: Cosmic rays, radioactivity and chemicals. Many sources have been researched for their mutagenic effect. This knowledge is useful to find out if animals and plants are at risc.

Examples of mutations in bees. Drones carry sometimes the mutation *white eyes*. They are blind because no pigment protects the light sensitive cells in the retina. The colour alleles for body colour are mutations of a single gene, e.g. yellow, brown, grey etc. Mutations are often recessive and cannot be seen in diploid organisms because dominant genes hide them. The haploid drone cannot hide any gene, and drones with mutated genes with negative effects will not mate a queen.

Mutated genes without effect in the drones, e.g. stinging behaviour, or pollen collecting mechanisms, are not sorted out through the drones. There are always some genes with negative effects in queens and workers. Together they are called the *genetic burden*. Breeders can find some of the negative genes through crossing; in fact, one of the reasons of noting your observations is that you might find out if something unusual might be inherited.

Vitality is a property, which is much desired. Vital colonies are strong, healthy and their production of larvae is big. Vitality is caused by the absence of bad genes and the preponderance of good genes. A single gene does not cause it.

Inbreeding and heterosis

Through inbreeding many genes are made homozygotic. This has often a severe negative effect because dominant ones do not hide bad genes.

If unrelated bee colonies with many homozygotic gene pairs are crossed many of the gene pairs become heterozygotic. The heterozygotic colonies are often strong, healthy and good in many ways. This phenomenon is called *heterosis*. The bees now have two different alleles in many places and such gene pairs produce two slightly different proteins, which make certain chemical processes in the cells more effective. Another fact is that good dominant ones mask many bad recessive genes.

We often see heterosis in crossings between two bee races, two pig races, and two pigeon races and in crossings between two inbred lines of maize or other plants. Heterosis is so effective that it is used commercially.

The first crossing has great effect; in the second generation the effect is lesser and dwindles in the following generations. We cannot explain why. The breeders who use heterosis must maintain the inbred lines and make new crossings every year. Some breeders of bees use that effect.

Nature and nurture

Every organism is a result of genetic (nature) and environmental factors (nurture). Which factor is the most important can only be found by examining every gene or character apart.

The history of the Western Nations during the last century shows the effect of the environment. About 1900 the average life expectancy was low (50 years for men, 53 years for women in Denmark). Now the average life expectancy is 72 years for men, 78 years for women. About 1900 13% of the newborn died in their fist year – now: 0,8%.

One hundred years ago the heating of many houses was insufficient and insulation was hardly in use. In winter people froze and the cold made the rooms humid. Because of that many people got sick. The food was scarce in many families and in springtime fresh vegetables with vitamins did hardly exist. Also the food problems made diseases common. Hungry people are easy targets for infectious diseases. The hygiene was bad and many bacteria and fungi could kill people.

The same environmental factors work in the bee hive. The hive must be well insulated and of good quality, the modern plastic hives fulfill these requirements. The hives must stand where the ground is dry and well drained, the wind weak, and the microclimate so warm that the bees fly out early in the winter and early in the morning. There must always be food enough, when this is not the case, then you have to feed the bees. Old combs, which have been used for larvae and storage of honey, should never go back to the hives because they contain spores of microorganisms, which could infect the larvae and bees.

Environment is everything which influence the individual from fertilization to death. Beekeepers can give their bees a fine environment and so can the bees themselves. They clean the cells and the interior of the hive, close openings with propolis, place their larvae where they easily can keep warm, regulate the temperature etc. The behaviour of the bees varies and the beekeeper can see if the bees do a good work. This is also used for selection of the best bees.

Some people use so much energy in learning

genetics that they forget to keep a keen eye on the environment. However, improving the environment is often the fastest way of improving the beekeeping.

Even the best genotypes are useless in a bad environment.

The relative importance of genes and environment

In breeding you must know if a character is heritable or caused by the environment, or if both influence it. Examples:

The development of wings. Without genes for wing formation the bee gets no wings. However, if Varroa has transferred *Deformed Wing Virus* to the larva the wings will become functionless and small. That virus is an environmental factor, which destroys the wings.

Queen or worker? The fertilized eggs contain all of the genes, which control the development of the larva to a queen or a worker. The cell form and the food determine the outcome of the egg. Here the influence of the environment on the genes is clear.

Nosema disease is caused by a fungus, which can kill the larvae and the bees. Some bees are very tolerant to Nosema, others are less tolerant, and still others get easily sick when they get Nosema spores. Danish breeders find bees without Nosema spores and use them in their breeding programme, see page 42. However, it is not known if these bees are without Nosema because of genes or because of strict hygiene.

The disease can hit even Nosema tolerant bees. If the bees winter in a humid hive, starve, or are disturbed they can be sick. The disturbing can be caused by branches beating the hives through a winter storm causing the bees to defecate inside the hive. Nosema spores from the feces infect the bees and might kill them. Beekeepers are able to control these environmental factors.

The Nosema story shows that beekeepers can improve both heritage and environment. The same holds true for the honey harvest, swarming and other factors. Every character is examined separately.

The influence of the environment is examined by keeping bees with a uniform heritage in different environments. In breeding you cannot get closer to uniformity than comparing sister queens from a line of bees with the same characters e.g. high honey yield, low swarming, peaceful, Nosema tolerance, and early development. The colonies are kept in bee yards with different environmental factors. If the characters remain unchanged or nearly so the heritage is most important. If they vary according to the environment it must be of greatest importance. This is seen in the results of "The Danish Beekeeper Association" control of queens from breeders, see chapter 7.

Heritability

The word *heritability* describes how much of a character is due to heritage, and how much is due to the environment. You often use heritability to judge the effects on characters, when you don't know the genes, for instance when a character is controlled by additive genes (honey production, disease tolerance etc.).

The easiest way of finding heritability is to see which characters vary and which don't by following what happens in the same bee family from one year to the next. As long as they have the same queen their genes are the same, variability must be caused by environmental factors. This way is not exact, but you get a good feeling of the heritability of the single characters. It is important to know that in breeding. High heritability means that you can make bees better using the genetic breeding methods. If environmental factors are the most important, you must work on them instead of using genetic methods.

14. Bee races and evolution

Before the beekeepers began moving the honey bee around it was living in Africa, Europe until the Ural Mountains, and the Middle East. In India and South East Asia 8 other bee species are found, e.g. *Apis cerana*, the dwarf *Apis floreae* and the giant *Apis dorsata*.

Our honey bee is called *Apis mellifera* in latin. *Apis* means bee, and *mellifera* honey-bearer. The latin names are used worldwide to avoid misunderstandings.

During the last centuries the honey bee has been taken to nearly every part of the world where it can live. European colonists have introduced it to North and South America, Siberia, East Asia, Australia and New Zealand. In some places the bees thrive better than in most European countries, e.g. Australia.

The honey bee originated in Africa according to the new genetic research (*Honey Bee Genome Sequencing Consortium 2006*). Here the greatest genetic diversity in the world exists and this is common to the place of origin in most species of plants and animals. During the last Ice Age the honey bee has been eradicated from most of Europe. When it ended 10.000 years ago the honey bee colonised Europe from refuges in Southern Europe and Africa.

Apparently, the honey bee came to Eurasia in at

Fig. 111. The original range of the European and North African bee races. In Kaukasus caucasica lives, in Armenia armeniaca. In the Middle East meda and syriaca live. The map is based on material from de la Rúa et al. 2005.

least two waves from Africa. One came to Spain and went on to Western and Central Europe. It is called the M-group. M stands for *mellifera*. The other wave wandered through the Middle East and Turkey to south Eastern Europe. It is called the C-group (C = *carnica*). It comprises *carnica, ligustica, macedonica, cecropia* and probably *sicula*.

The honey bees of the Middle East, Turkey and Caucasus are distinct from the C-group and are called the O-group. The different honey bees of Africa are included in the A-group.

The map (fig. 111) shows which races belong to the groups. A race is a population of bees which can be distinguished from other populations by morphological characters, which means that they can be seen on the bee. To day the races can be distinguished genetically, too. In most cases the results found by morphology and genetics are the same.

The races were isolated geographically until the bee keepers began moving bees and queens around. The result was a mixing of races. Some of them disappeared.

The *Nordic bee* is dark brown and lived in Western and Northern Europe. The name of the race is *Apis mellifera mellifera*. The name of the race has three parts, the first is the name of the genus, the second that of the species, and the last one the name of the race. The name of the *Italian bee* is *Apis mellifera ligustica*, the *Carnica bee* from Central and South Eastern Europe is called *Apis mellifera carnica*. Talking about races, we use only the racial name, e.g. *ligustica, carnica*.

The map of the distribution of races shows the original state insofar it is known. Some of the races have been distributed by man in great areas. The original bee in England was *mellifera*, but in the years about 1914 it was nearly eradicated by the tracheal mite. In the years about 1920 the *ligustica* bee was imported from Liguria, a region around Genova in North Italy. It was tolerant to the mites, gentle and a good collector of honey, too. It is common in England today.

In Germany the original *mellifera* was exchanged with *carnica* nearly everywhere in the years after 1950 because of its gentleness and its general good characters.

A geographical race has adapted to the special conditions of life in its area. Certain genes are common to all of the members of a race. It can be seen in the ability to survive and in the behaviour. An example is the original bee from Caucasus, *caucasia*. It collects propolis eagerly and overflows the hive with that stuff, while the Egyptian bee, *lamarckii*, don't collect propolis at all. *Carnica* is gentle while many African races are extremely aggressive because they are exposed to many more predators than European bees, e.g. honey badgers and other mammals, bee eaters and other insectivorous birds, certain reptiles and toads, other insects and man. In Gambia in Western Africa the beekeepers harvest only honey at night and they wear heavy protective clothing. If they harvest the honey in daytime the bees continue their attacks on every living thing until sunset.

When the African bee race, *scutellata*, was imported to Brazil in 1957 as an experiment it crossed with the European honey bees which had been imported several hundred years earlier. The result of the crossing was the *africanized bee* which is ill famed because of its aggressivity. In a few seconds a guard bee can call the other guard bees to attack and the other colonies in the bee yard attack, too. They pursue people and domestic animals far away and sting them. People can get more than a thousand stings. However, these bees yield much honey and are more or less tolerant to *Varroa*. In Brazil combat of Varroa is not seen as necessary.

Since 1957 the africanized bee has spread to most of Latin America. In 1990 it arrived at the USA where it is still spreading. It attacks European bee colonies, kill their queen, bring their own, and take the colony in possession. Many counter measures have been tried but the future outcome of the fight against it is insecure.

The origin of races

The honey bees are expanding their geographic range, just like all species of plants and animals. If the conditions of life in the new area fits the bees they establish new colonies there. However, the bees in the new place bring only a limited part of the genes of the population with them. If the area is cut off from the original area by e.g. climatic change or by rising of the sea level the new population cannot get genes from the original population. After that, there exist two genetically different populations of bees.

Isolation of bee populations have often happened. Four or five thousands years ago Sahara was a savannah with a rich life and honey bees in many places. Then the climate changed and Sahara became a desert. The bees survived in isolated areas (oasises), and the North African bee populations were effectively separated from tropical Africa. The isolated bees evolved in new races *(sahariensis, lamarckii, intermissa)*.

The sea all over the world rose 30 meters in 1500 years about 7000 to 8000 years ago. Much land was drowned and divided in islands (fig. 88).

When a small population of honey bees are left without contact with other bees the gene pool can change by chance. It is called *genetic drift* (fig. 87) and works in this way: If there are, say 10 colonies, in an island they will reproduce as usual. The colony with the biggest production of drones will mate more queens than the other ones. Then the genes of that family will become more common, and if the offspring also maintains a high rate of drone production their genes will dominate even more.

Accidents can influence the result. Some queens are eaten by birds and their genes are not reproduced. Some colonies live in a place where they die from hunger while colonies in other places survive. In this way genes are sorted by chance. If the bees of the isolated place get more room they will expand. A new race with its own genes or gene combinations can be started in that way. On the other hand: If beekeepers introduce another bee race to an isolated population the genes are mixed and the old population often loses its special characters.

The natural dispersion of honey bees is small because it is done by swarms which only move a short distance, maximally few kilometres. Today colonies are moved by the beekeepers, e.g. from Australia to the almond plantations in California, as package bees to the Northern USA, or across Russia. Queens are sent from Hawaii to many other places in the world, and similar transports are happening in many other places. Swarms are sometimes found on ships and presumably some parts of the USA have got their africanized bees in that way. In Australia many ports have swarm traps with pheromones which should attract swarms from ships, it is hoped.

All this travel demand much of the bees. They must be able to adapt to quite new environments. If they can't they perish. How different the environments are can be seen in Europe. In Sweden the winter is long and the bees must collect honey for 7 or 8 months. During the winter the bees must stay inside the hive. In the Mediterranean region the rains mainly falls in winter and the flowering occurs mainly in that period. The summer is hot and dry without wild flowers and the bees need honey to come through that period. The bees must also cope with diseases and predators which are not found in the original place.

How do the bees cope with all that? Every swarm is genetically different from other swarms because of meiosis and cross-over. The reproduction of bees will make new combinations of genes and that is a major key in the adaptability of bees. If some colonies with inborn rhytms fit for fast development in spring and no resistance to tracheal mites are crossed with bees with slow development in spring and resistance to tracheal mites, then some of the offspring will combine fast development in spring and resistance. If that combination fits to the environment it will continue to exist because the other combinations will be eradicated by the environment. In this way new differences between races can develop. This process is called *acclimatization*.

The breeders use the same method as nature. They combine colonies with different useful genes and select the combinations they wish.

Nature is changing

Nature is changing through the times and the bees must adapt genetically to the new factors of the environment. The climate is changing: Since the end of the Ice Age 10.000 years ago Northern Europe has had two periods with polar climate, and periods with a warm climate, for instance during the Bronze Age. From about 1400 to 1700 A.D. the climate was cold and wet, the so called Little Ice Age. Later the climate became warmer. During the last 30 years the period of plant growth has become one month longer in Denmark.

When the climate changes the vegetation changes too. New plants with other flowering times arrive and the rhytm of the year changes. When agriculture came to Europe 5000 to 7000 years

RACES	fertility	eager to collect	disease resistance of brood	disease resistance of bees	swarming tendency	long life	ability to fly	weather resist.	sense of finding	honey far away from brood	eager to build	gentle	staying on comb	use of propolis	build wildly	orientation
Buckfast	+4	+4	+3	+5	+6	+2	+2	+5	+5	+6	+6	+6	+5	-5	-5	-1
Ligustica	+3 +4	+2 +3	+3 +4	+3 +3	+3 +1	+1 +2	+1 +2	+1 +2	+3 +3	+4 +4	+4 +4	+4 +5	+3 +4	+2 +1	+1 -1	-2 -1
Carnica	+2 +2	+3 +4	+5 +5	+2 +3	-5 -6	+4 +4	+2 +2	+3 +4	+2 +2	-1 +1	-2 +1	+6 +6	+6 +6	+2 +1	+3 +1	+3 +3
Cecropia	+2 +5	+3 +4	+3 +3	+2 +3	+1 +5	+3 +4	+2 +2	+3 +4	+2 +2	+1 +4	-1 +4	+4 +5	+4 +5	+2 -1	+3 -1	+2 +2
Caucasica	+1 +3	+1 +2	+1 +1	+1 +1	+1 -1	+1 +1	+1 +1	+1 +2	+1 +1	-6 -1	-6 -1	+6 +6	+6 +6	+6 +4	+6 +4	+1 +1
Intermissa	+1/+3 +4	+4 +5	-4 -4	-3 -1	-4 -5/+3	+6 +6	+6 +6	+6 +6	+6 +6	+1 +3	+5 +5	-6 -1/+2	-6 -1	+6 +5	+6 +5	+3 +3
Mellifica mellifica	+1/+3 +4/+5	+5 +6	-3 -1	-3 -1	-4/=3 -5/+3	+6 +6	+6 +6	+6 +6	+6 +6	+2 +3	+6 +6	-5 -1/+2	-5 -2	+6 +4	+6 +4	+3 +3
Mellifica lehzeni	+2 +2	+5 +6	-3 -1	-1 +1	-6 -6	+6 +6	+6 +6	+6 +6	+6 +6	+2 +3	+6 +6	-5 -1	-5 -2	+6 +4	+6 +4	+3 +3
Fasciata	+1 +3	+2 +3	+2 +3	+2 ++3	-1 +2/+3	-1 +1	-6 -5	-6 +1	+3 +4	+1 +4	-1 +3	-5 -1	-5 -1	-6 -4	-6 -4	+6 +6
Cypria	+1 +3/+5	+2 +5	+2 +3	+2 +3	-1 -4/+3	+2 +3	+2 +3	+3 +5	+4 +5	-1 +3	-1 +3	-5 -1/=2	-5 -1	+1 -1	-6 -2	+6 +6
Central anatolica	+1 +3/+5	+6 +6	+2 +3	+3 +4	+2 +5	+6 +6	+6 +6	+5 +5	+3 +4	+1 +3	+2 +2	-1 +2	-1 +2	+3 +2	+3 +2	+3 +3
Sahariensis	+1 +5/+6	+6 +6	+3 +3	+3 +3	+3 +2/+4	+4 +5	+4 +5	-3 +5	+6 +6	-1 +4	+1 +4	+2 -1/+2	-6 --1	+2 +1	+3 +4	+4 +3

Fig. 112. Brother Adam's scheme over genetically influenced properties. The properties of the Buckfast bees are described in the upper line with scores: +6 is the best score, -6 the worst. + are positive properties, - are negative ones. The properties of natural races are shown as the first number in every rectangle. The second number is the result of the offspring F1 after crossing with the Buckfast bee. The third number is the result for the F2 generation. Compare the map fig. 111.

ago it changed the landscape. Many forests were changed to fields or grassland and domestic animals did grass in the forest. In some areas the heaths took the place of the forests and their dominant plant, the heather, became and important plant for bees. To use it the bees had to change their rhytm of the year. The way agriculture cultivates the soil is changing in these years. The fields have grown and now they are big areas with monoculture and no weeds because of herbicides. After the harvest the fields are sown again. There is no room for wild plants.

The bees must follow the changes or perish. They do it in the same way as they make new races.

The fate of the races

Two hundred years ago there were many ecotypes with their own genetic peculiarities. An ecotype is a population within a race. It has adapted to special ecological conditions of life. The ecotypes began to disappear when railways became common after 1850. Now it became easy to send queens across the frontiers. The German railways had a special service for the bee keepers of Germany. The could send their straw hives by train to the enormous heath around Lüneburg. The service was widely used; there were special wagons for bees and special bee yards on the stations where the railway personel placed the hives so that the bees could collect the prized heather honey.

When good roads and lorries became common in Europe after 1950 the wandering of bee colonies became widespread. The import and export of queens rose. Because of that the local races disappeared by mixing with other bee races in most places in Europe. The *ligurian bee* which is a local variety of *ligustica* cannot be found in Europe, but it is still alive in Kangaroo Island South of Australia. It was imported to that island in 1885 from Liguria and no other bees have come to that island.

Ecotypes of the Nordic bee, *mellifera*, exist in different places of Sweden, Norway, Denmark (the island Læsø), Scotland and elsewhere, See fig. 114.

How to use the races
Some people think that you ought to keep pure races in the area where they belong because they are best adapted to the natural environment. This need not to be true. The bees of the present time have to adapt both to the nature, the local land use, and the local way of beekeeping. From about 1960 to 1990 the Danish farmers had very big areas of colza. The variety they used had its flowering time from mid June to mid July. The greater part of the honey harvest came from these flowers. The bees were able to use it because they were at the peak in numbers of workers. After about 1990 the farmers changed the variety of colza. It flowers in the beginning of May when the number of workers is still low. Now, we need a bee with early development, but we have not got it yet.

The bees must adapt to rapid changes in the environment, and they do it. The differences in gene combinations is the base of natural selection which means that the most fit bees survive while the others die. The process takes some time.

Even if the races or ecotypes have disappeared from most of Europe their genes are still present. Because of that an old race can be recreated through crossings and selection. It has no purpose in itself, the races should be used in another way. What characterizes a race is that all of its members have certain combinations of genes in common and sometimes specific genes which only are found in that race.

Brother Adam and other breeders have collected bees and informations about them in North Africa, Africa South of Sahara, the Middle East, Turkey and Mediterranean islands and isolated land areas. In that way they found bee populations and races with useful characters e.g. low swarming tendency, good rhythms of the year, low food consume during winter etc. Queens of the useful bees have been brought to Europe and crossed with other bees after a precise planning. In that way many fine characters have been brought together in Broder Adam's Buckfast bee. His work shows why the different races ought to be maintained.

Breeding may have dangerous consequences. The aim of breeding is to create a bee which has almost the same characters through the generations. If a certain bee race obtains great success it will be used by most beekeepers. In that way many genes disappear which could be useful when the environment changes or new diseases arrive. The consequences could also be that certain genes which protect the bees against common, but not dangerous diseases, disappear by accident. Then the disease becomes dangerous.

This risk is well known by the plant breeders. They fight it by making *banks of genes* where old races and original wild types are kept alive. The banks deliver material for crossing with the commercial types to renew them. Gene banks for plants are collections of seeds and plants. The solution for bees is to keep certain races and populations alive in restricted areas where no other bee races are allowed to live. It works e.g. in Norway. In Denmark the island Læsø has a refuge for the special population of the Nordic bee, *mellifera*.

Another reason why the races and special populations must be maintained is *heterosis*, the good effect you might obtain by the crossing of two different races.

Morphological characters of races
A condition for work with the races is to be able to recognize them. Ruttner (1986) has made a thorough analysis of the morphological characters and built up a reference collection of the races in Oberursel (Austria). According to him there are 24 or 25 races.

Among the morphological characters are colour, hairs, tongue length and wing venation.

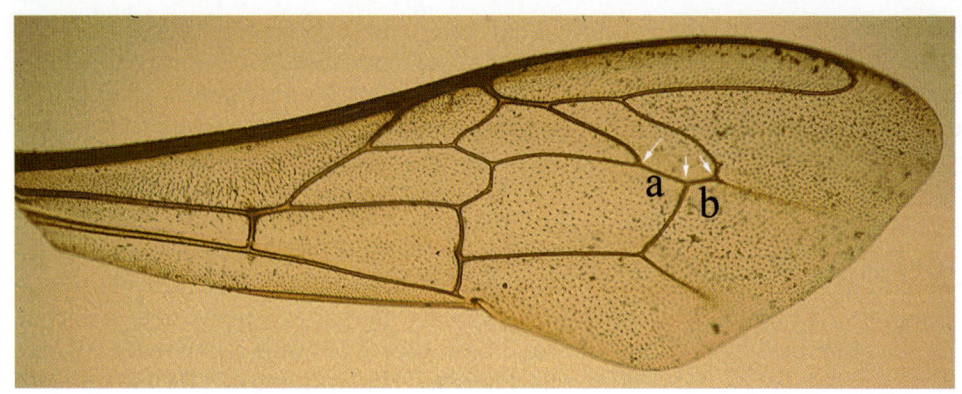

Fig. 113. Forewing of a worker bee. The white arrows are inside the cubital cell. a and b are the two lengths you measure. You divide the bigger number with the smaller one. In this way you obtain the cubital index.

Some people think that a yellow bee is a *ligustica* and that the colour is a guarantee of racial purity. That is not the case. Because of the mixing of races the colour can be combined with a lot of good and bad characters. It is often seen in colonies where the queen has mated with different drones in nature. Some of the drones had genes for yellow, brown or other colours, accordingly the workers have different colours. However, the colour can be used as a marker for a race or a line if you breed such that all of the bees have the same colour. Then another colour tells about wrong matings.

In some cases the colour can be linked to certain good characters. If the gene for colour and an useful other character are so close to each other on the same chromosome that they rarely are separated during cross-over in meiosis, then you can use the colour as a marker for the useful gene.

Another useful morphological character is the cubital index (fig. 113). It is found by examining the forewing of 30 workers from a colony. You measure the length of two wing nerves along the third cubital cell (fig. 113). After that you divide the greatest number by the smallest and thus gets the cubital index. The examination is done in this way: You cut off half of the forewing and place it under a microscope with a ruler, and makes the measures.

The races have different cubital indices, but they overlap. Therefore you need to measure 30 wings to calculate the average index.

Genetic examination of races

The Danish scientist, *Bo Vest Pedersen*, has studied genes of bees from many races and ecotypes. The results are summarized in fig. 114 in the form of a pedigree. The result was, that the *mellifera* race is very different from the other races (*carnica, anatolica, sahariensis* and a number of African races). Apparently, the antecedents of the honey bee were divided in two branches during its evolutionary history. One of the groups evolved into *mellifera mellifera*, while the other branch divided in three groups: The African races (A-lineage) and the Mediterranean lineages C and O. C comprises *carnica, ligustica* and others, while O comprises *anatolica, caucasica* and others.

The result came from the examination of genes of the mitochondria. In a gene for a respiration enzyme *(Cytochrome oxidase I)* 900 pairs of bases from different races were sequented. The gene of *mellifera* had 15 pairs of bases which were different from the genes of *ligustica, carnica, cypria, anatolica* and *sahariensis*.

A similar difference was found in the region of the mitochondrial chromosome which is placed between the genes for cytochromoxidase I and II.

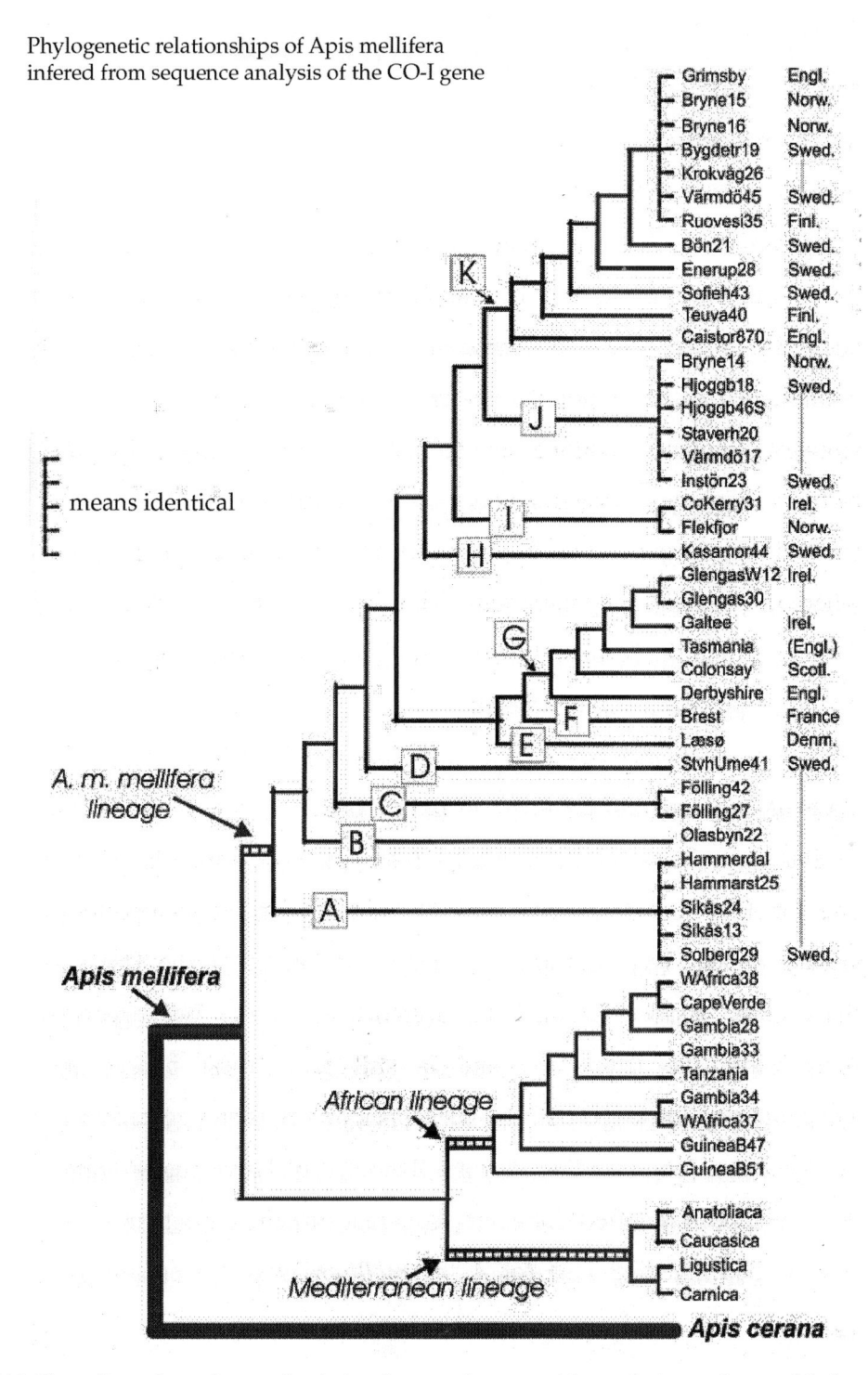

Fig. 114. The pedigree shows the genetic relations between the geographic populations and races of the bees. The existing races and populations are seen to the right. If you follow a single population from the right towards the left you see the forefathers. The African races and those of the Mediterranean are seen in the lowest part of the figure. They have evolved independently. The Nordic bee (m. mellifera) fills up the rest of the pedigree. Every forking means that a population has divided in two genetically different populations. The closer the bees are in the pedigree the closer relatives they are. (Bo Vest Pedersen 2005).

14. BEE RACES AND EVOLUTION

15. Breeding

In the breeding of bees you try to combine many useful characters in a group of bees, and get rid of the bad ones. The aim is to make a race or a stock. However, it is artificial but fits the purpose. Every, or nearly every member of the race or stock has all the selected characters and because of that you are sure that the offspring has the same characters if the matings take place within the race.

There is no principal difference in methods when Nature makes a race, and when man does it. You start with the genetical variation, get acquainted with it, combine characters through crossings, select the fittest, and isolate the unmated queens so that unwanted matings don't take place. Breeding works together with Nature, not against it. Indeed, the work is big and difficult. In this chapter the methods of the breeders are described.

The breeders' work can never stop because the only way to maintain the desired characters is to control the breeding meticulously. If you lose control for one or two years all the work is lost. The bees don't know that they are something special and mate with the drones there are in the local area. The bees behave like dogs; if the dogs could mate freely all of the races would be mixed in a couple of generations.

The Buckfast bee

During 73 years, from 1919 to 1992, the German monk, Brother Adam (1898 – 1996) was leader of the beekeeping in Buckfast Abbey in England. He began in the years when 90% of the English bees died from the attack of the Tracheal Mite. In 1920 he crossed the Italian bee (*ligustica var. ligurica*) with the English bee of the Abbey. It belonged to the Nordic bee race (*mellifera mellifera*). He obtained bee colonies which were resistant to the Tracheal Mite and had other useful characters. According to Brother Adam about 25 years passed until all the colonies were fully resistant (1947). In this way the Buckfast race started. It is a product of human ingenuity like the races of cattle.

Brother Adam continued his work with the race and made it genetically stable. Already in 1930 he made his first planned crossings with French bees. According to his pedigrees he used Greek bees in his crossings in the 1950'ies. Anatolian and Egyptian bees were used in the 1960'ies. During the years Brother Adam imported bees from many countries to his experiments after having studied them on location. He continued this work until he left the Abbey in 1992. Buckfast Abbey tries to continue the work using his methods.

The work continues, too, in Sweden, Denmark, Germany and other countries. Here the beekeepers have imported many stocks from Buckfast Abbey.

Combination Breeding, an overview

The Danish Buckfast breeders are organized in an association (Avlerringen). They call themselves *Combination breeders.* They continue the work of Brother Adam after his methods. However, the breeding can never stop and the breeders must evolve the bee as described below.

Every year the combination breeders decide which stocks of bees shall be used for delivery of drones on the mating stations. It is published, and now the individual breeder decides where his queens are going to be mated. It depends on the stock to which the queens belong because the individual stocks are not of equal value in the crossings.

What is a stock? It is a group of colonies which descend from one queen which had a successful mating. The stock has a certain combination of useful characters which continue through the generations to come if the breeders use the right methods. The Buckfast race contains many stocks which must be paired with each other to maintain the race.

It is necessary to use many *unrelated* stocks in the work. Otherwise the race is destroyed by inbreeding and diminishing numbers of sex alleles. *Unrelated* means that the two stocks of a mating at most have a great grandmother in common.

Maintenance and crossings of stocks is not enough to keep the vitality of the colonies. That is because you make a selection of queens to find the best, while you take the not-so-goods off the breeding. In this way you impoverish the stock by sorting out many genes. Because of that you need to add "fresh blood" to the stock which makes the genetic variation bigger. It is done by crossing selected stocks from other races with selected stocks of Buckfast thus creating new stocks. These are examined through some generations of breeding until you know their performance. This method is called *combination breeding* and is described below.

Crossing of races can be used to make *heterosis* (se page 73).

Selection

Breeding begins with controlled matings where selected queens mate with selected drones in a mating station or by instrumental insemination. You can make thousands of queens if you have enough of cell builders and nucs, and time to take care of them. Exactly here you have the greatest difficulty. A breeder of wheat or barley is able to work with millions of plants during the selection process while the breeder of queens at best only can work with few hundreds of sister queens.

Brother Adam worked with so many series of queens that he could take 80% of them away by inspection when they emerged from their cells. He used the colour although it is insignificant by itself. The selected 20% were placed in nucs and when they had produced their first brood he selected again. About half of them were not used for breeding but were used in colonies for honey production etc. The other half were used for breeding. When their colonies had grown big and had functioned for some time he selected queens for further breeding.

The following years new generations replaced the former and Brother Adam continued to select the best queens until he knew that the offspring had the same characters as the parents. It takes 7 or 8 years to create a new stock in this way. It must be better than the existing stocks, if not, it is abandoned.

A queen which is used for breeding is selected by the characters of her progenitors, her own colony and her offspring through two years. Very good queens are kept alive as long time as possible, that is 4 or 5 years.

The Danish combination breeders have decided that a queen for breeding must be judged upon at least 30 sister queens in colonies of their own. Accordingly you have to produce many more through grafting.

The selection is thorough. It demands pedigrees, score sheets and description of the heritage for 5 or 6 queens at least.

Today the Buckfast method is used by breeders of ligustica, carnica and other bee races and stocks.

Crossing strategy

The chapters on genetics describe crossings and their outcome. You must make a strategy to be able to use the genetics in breeding. Here you are an example.

The crossing of brown bees (yy, recessive) and yellow bees (YY, dominant) show this:

P	YY x yy	Two genotypes, two phenotypes
F1	Yy	One genotype, one phenotype (two F1 are crossed)
F2	1YY:2Yy:1yy	Three genotypes, two phenotypes

In the P-generation the crossing is made. In the F1-generation all are alike, heterozygotes, because the P-generation has two different homozygotes. In the F2-generation a segregation occurs because two F1 were crossed.

This example demonstrates what happens during crossing, even when the genes are unknown. If F1 is uniform we know that the genes of P are as shown. Then there will be a segregation in F2. Now it is time to look after bees with desired characters. If a character is controlled by recessive genes it is easy to find; about 25% of the queens are yy. Further crossings between yy males and females will not result in further segregation.

If the character is determined by the dominant gene Y you must know that you cannot distinguish YY from Yy. (If the genes control colour you can see it on the drones; offspring of Yy queens has yellow (Y) and brown (y) drones, while the drones of YY are uniform). If you cannot use the drones you select a number of queens from F2. All of them are inseminated with sperm from drones from a yy-queen. Thus you perform an analysis test. The F3-queens whose mothers are GG will make G-eggs, and by insemination with g-sperm all of the offspring will be Gg. Result: No segregation.

50% of the F3-queens whose mothers are Yy will produce Y-eggs, the other 50% y-eggs. By insemination with y-sperm 50% of the offspring becomes Yy and 50%, yy. Result: Segregation.

After that we know which F2 queens are Yy and which are YY. Next time you use their larvae for grafting you are sure which of them are homozygotic YY.

The same method is used to find unrelated queens which deliver Y-drones for insemination or on mating stations.

If you want to find yy in F2 after a crossing YY x yy in P you need more then 4 colonies because the chance of getting yy is 25% *on the average*. If you want to find a double recessive *ffss* in a crossing with two pairs of genes as in the example with hygienic bees (fig. 97, 98) there is only 1 chance out of 16 to find it. In this case 16 colonies are not enough, you need a much bigger number.

If you look at the heritage when three pairs of genes are involved (fig. 100) there is 1 chance out of 64 to find the triple recessive homozygote. The work becomes impossible. *Accordingly the common strategy is that you work with one character at a time. When you know that the queens are homozygotic in one character you proceed to the next character.*

If a character is based on *additive genes* you get a uniform F1 and a segregation in F2 by crossing two F1's. As you don't know the number of genes you cannot calculate how many colonies you need. You make so many queens as possible and examine so many of them as you can. You can have luck and get queens with uniform offspring during few generations. However, in other cases you need to use many years of breeding to obtain the desired goal.

You need to know if a character is influenced by the environment and how much. The easiest way to obtain some knowledge about it is to compare the characters in a colony with the same queen through two or three years. (page 74, Heritability).

Testing

After the testing of the queens by the breeder their daughters should be tested in bee yards of different beekeepers. As the environment is very different you get a test of their genetic outfit in different conditions.

The Danish Beekeepers Association cooperates with the breeders about testing. There are two test programmes, one for queens sold to beekeepers for honey production (1991 to 2008), the other for disease tolerance in queens used for breeding (2001-2008). The testing continues.

The tests of the first programme has demonstrated a very clear improvement of the genetic outfit. The aims of the breeders have been to improve the honey production, make the tendency for swarming low, make the bees gentle and quiet.

The second programme tests the bees for hygienic behaviour, and the improvement has been great from 2001 to 2007. Hygienic bees are tolerant to American Foulbrood and other infectious diseases, and perhaps Varroa. The testing shows also that environmental factors can influence hygienic behaviour, for instance the weather, the honey flow, the time of the year, and the number of bees in the hive. Because of that the test for hygienic behaviour is made on exactly the same date everywhere in Denmark. A combination of unlucky factors made the results bad in 2006. In 2007 the bees were back on the improvement track.

The Danish Beekeepers Association recommends its members to buy their queens from breeders who participate in the testing programmes.

Maintenance of stocks
To make a useful stock is a huge work and therefore it must be maintained so long time as possible.
Four factors are important:

1. The queens (and the workers) must be homozygotic for the most important characters because segregation should be avoided.

2. Bad characters must be removed by selection.

3. Certain useful characters of the stock, for instance low tendency to swarm and gentleness, should remain combined.

4. There must be so many sex alleles in the stock that the maximal non-hatching eggs don't exceed 10% to 15%, rarely more.

A basic method is controlled matings where you make inbreeding in one generation, and outcrossing with an unrelated stock in the next. The inbreeding is shown in fig. 115. You take two sister queens from A. Both get a colony of their own. One of the sisters, B, deliver the drones which mate with the daughters of the queen of C. Genetically spoken it is a *backcross* because you mate offspring with its parents.

There are other types of inbreeding, but the one described is the best in maintenance because the genetic damages are kept at a minimum.

In the crossing page 83 you see that the offspring of the crossing Aa x Aa gives 1AA:2Aa:1aa. This generation has 50% homozygotes while the P generation had none. By close inbreeding as just described there are only two alleles, A and a. Therefore, homozygoty is readily obtained, and that is what is desired.

But undesired recessive genes get homozygotic, too. Selection must be used and it must be done through some generations.

By the mating of a queen with the drones of her sister the number of sex alleles diminish rapidly. If the queen of A has, say, 12 different sex alleles in her spermatheca, one of these is paired with one of her own two sex alleles when she produces an egg for a queen cell. If you use two sister queens of her offspring they will at most have 4 sex alleles, and the result is 25% unhatched eggs

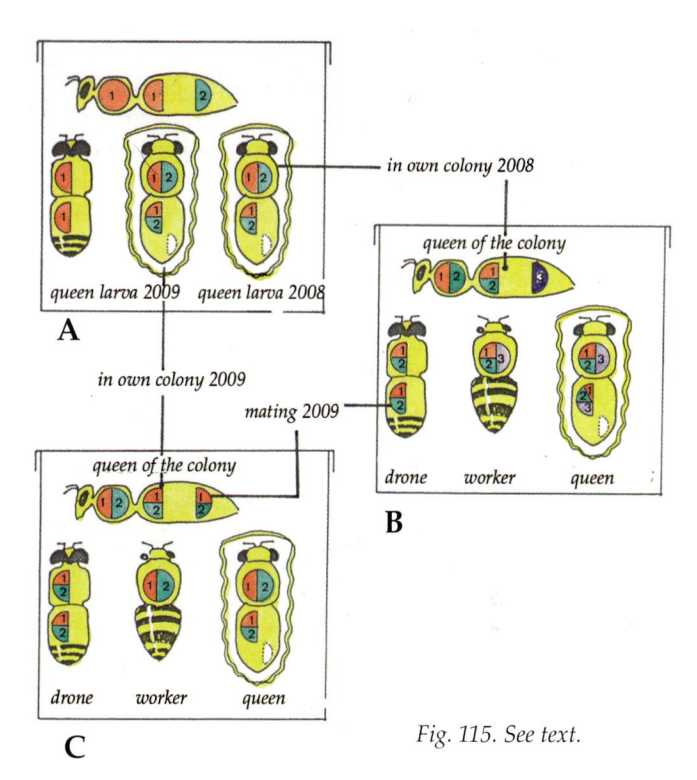

Fig. 115. See text.

in the mating between B and C. The minimum is two sex alleles which gives 50% unhatched eggs. This is the most important reason why inbreeding is not performed in the next generation. Instead, you cross with an unrelated (or little related) stock which has almost the same useful combination of genes as the stock you want to maintain.

After this outcrossing many queens are produced for selection. The percentage of unhatched eggs are measured to be sure of the number of sex alleles.

It is impossible to make strict rules for when you should make inbreeding or outcrossing. Both methods must be used. Experienced breeders use both their knowledge and intuition in their selection of stock for mating.

Experienced card players know that two mates can have 13 cards each which could be combined in such a way that you win. But the two players could have sets of cards which cannot combine, and you loose. You could look on the many stocks in the same ways. Some of them combine well while other stocks do it badly. This is one of the reasons why pedigrees and score sheets are important in the planning of breeding.

15. BREEDING 85

Stocks for drone production

A stock for drone production must be genetically stable because it is used for the fertilization of many queens. Because you need many different sex alleles you cannot use a single colony as drone source. Instead you use a number of sister queens from a queen which has been tested through two years and is genetically stable. She has been mated with drones on a mating station and because of that she probably has sperm with many different sex alleles in her spermathece. Her daughters are uniform concerning the important characters (homozygotic). This is obtained by crossing of stocks which have the desired characters if breeding and selection is used before you choose the sister queens for the mating station. At least 6 sister queens are used there but 12 or more is much better.

The drone stock is more important than the stocks which deliver the queens, simply because a stable stock of drone deliverers fertilizes many queens from different stocks.

Closed populations

If you produce queens for sale to the beekeepers you must be sure that the quality is uniform or nearly so. Then your customers will be happy and return. A condition for that is that you have genetically stable bees.

You start by collecting 50 or more colonies with uniform and stable characters. When you have enough you close the populations which means that all the matings are done by instrumental insemination and that sperm and new queens come from your stock. Such a population can be used in many years before it needs supplying genes from new stocks.

The breeding can be further stabilized by using sperm form several hundred drones. It is mixed completely through centrifugation, a process called homogenization. A portion of this sperm will have a sufficient number of sex alleles and sperm cells from all of the drones. This makes the offspring more uniform.

Queens are produced in that way in countries or areas without islands, for instance in USA, Poland, Germany.

16. A German breeding system

Kaspar Bienefeld and his collaborators of the Institute for Bee Research, Hohen Neuendorf, Germany have developed a mating system founded on queens with a pedigree. The queen breeders can choose the stations where their queens obtain the best matings. It is possible to predict how much the chosen traits will be improved.

The system is founded on carnica. By sticking to a pure race you get no big heterosis effects with unwanted segregations in the next generations.

In Europe the system is used in Germany, Austria, Norway and other countries. In Norway the Norwegian Beekeepers' Association will use it on carnica and the Norwegian mellifera. It will be done in close collaboration with Kaspar Bienefeld and his collaborators.

In Germany the database contains results from tests of 100.000 carnica queens. About 6000 new queens and their pedigrees are registred in the database every year. Only the tests from the last 5 years are used. Then the base of comparison remains up to date and it is possible to measure the progress.

The database gets the pedigrees of every queen and the scores of the tests of her colony. The breeders have access to all of the datas if he has registered through his local beekeepers association.

What has the breeder to do? He goes to www.beebreed.eu. Here he logs in through the number of his association, his personal breeder number, the pedigree number and year of birth of the queen.

After that he has to choose either instrumental insemination or mating on a mating station with many sister queens just like it is done on the Danish Islands. In Germany many mating stations are placed in isolated valleys or on a small number of islands. The German mating stations are listed in the homepage.

When the mating station is chosen you write its data together with the informations you already have written. After that, you get the degree of kinship of your queen with the drones of the mating station. You also get a prediction of the progress for every trait. If the result looks bad you just try another mating station.

The degree of kinship is calculated by the database. It uses the informations from all of the pedigrees in it. The percentage goes form 0% to 100% kinship.

We know that inbreeding often is injurious partly because of the diminishing number of sex alleles, partly because noxious genes often become homozygous. Because of that it is useful to know the degree of kinship, the so called kinship coefficient. Some inbreeding might be useful especially if a certain crossing gives good results.

Breeding value is given in %. 100% means the average of all colonies through the last 5 years. If the honey harvest is 120% on the mating station and your queens come from colonies with 100% you can predict a future harvest somewhere between 100% and 120%.

Gentleness 100% means the degree of gentleness of all colonies through the last 5 years. It is calculated by using the scores of gentleness as in Denmark.

If a planned crossing predicts breeding value of 110% for honey harvest and 80% for gentleness you must choose if you prefer more honey and less gentleness, or rather would like another crossing.

Breeding value is the value of the genes of a certain trait. However, most traits are a product of the genes and the environment. In the system of Bienefeld the influence of the environment is neutralized, only the influence of the genes are measured.

As a rule the influence of the environment on a trait is found by observing how it varies with the environment. The enormous set of data in Germany shows how much the single traits vary. In this way the influence of the environment can be judged and subtracted from the breeding value.

Inbreeding might diminish the good influence of certain traits. Because of that the degree of inbreeding is calculated and subtracted when the breeding value is calculated.

The German Carnica breeders get a prediction of the mating result before mating or insemination. Accordingly, the breeder can choose the best mating. The system is based on a test of the mated queens and their colonies. It is done in a way which closely resembles the Danish evaluation method, which is described in this book. The queens get a pedigree number and their mating is recorded. The results of every mating, also the bad ones, go to the database and are used in the future calculations.

The system is always improved. Just now a new value of varroa resistance has been developed. It makes the possibility of breeding varroa resistant bees better.

17. The future

Breeding of bees has given fine results. The honey production has risen 2 or 3 times during the last 40 years. The bees swarm less, in most cases only when the beekeeper has forgotten to enlarge the space within the hive. Aggressive bees have been exchanged with gentle ones. The proportion of hygienic bees has become larger and is still in progress.

All this demonstrates that the methods of the breeders have worked well. The results are partly caused by the use of genetical knowledge, partly by the improvement of the environment.

Breeding is governed by wishes and need. Nature and agriculture are changing, and so is the climate. New pests appear, or the old ones change their virulence. The bees and the beekeepers must adapt to all that.

Now we know the genome, and hence the genes and the microsatellites of the bees. However, we also need to know the task of every gene, a work that will take many years. It is important to know the alleles of the genes, too, because much breeding consists in finding certain alleles and crossing them into the stocks with which you work. The bee geneticists must follow the progress and make new knowledge available to the beekeepers.

How do you know if your bees have the desired genes? There are many ways. It is still useful to describe the properties and note them in the pedigrees and score sheets. A new method is to find *marker genes*, which demonstrate their presence by certain peculiarities in the morphology, physiology or behaviour of the bees. That method has proved to be useful in the breeding of wheat and barley.

Another method is the use of microarrays which directly show if certain genes or alleles are present. Today, they are used in laboratories but hopefully they will become so cheap and easy to handle that breeders can use them. It would be a great help if we could find the presence of a gene within a couple of hours instead of using test crosses through a couple of generations.

Modern DNA techniques have proved useful in characterizing races and stocks. They can be used, too, to see if wrong matings have happened in the mating stations. New mutations can be found by these techniques.

Which are the desires of the breeders? Most important is tolerance or resistance against pests, for instance viruses, Varroa and other mites, the Small Hive Beetle etc. Hygienic bees are rather tolerant of American Foulbrood. The story of the tracheal mite in England shows that mites can be conquered by breeding.

The fight with pests will never stop. They will evolve further and might break the tolerance or resistance of the bees. Every breeder knows that. Yet, breeding is the best thing you could do because it does not poison the environment. This is contrary to the effect of pesticides, which should not be used except in an emergency. The pests evolve resistance towards pesticides and when it has occurred we have lost the medicine.

The breeding of bees is based on the principles of evolution, and so is the development of the pests. As we know how it occurs we can live with the problems because we know how to solve them.

Literature

Adam, Broder, 1982: Züchtung der Honigbiene. - Delta Verlag, St. Augustin. 144 p.

Adam, Broder, 1983: Auf der Suche nach den besten Bienenstämmen. - C. Koch Verlag, Oppenau. 157 p.

Adam, Broder, 1983: Avelsarbete med honungsbin. - Kristianstad. 16 s.

Bienefeld, Kaspar et al., 2007: Genetic evaluation in the honey bee considering queen and worker effects - A BLUP-Animal Model approach. - Apidologie 38, p.77-85.

Bienefeld, Kaspar et al., 2008: Noticeable Success in HoneyBee Selection After the Introduction of Genetic Evaluation Using BLUP. - American Bee Journal, vol. 148/8 p. 739-742.

Cobey, Susan, 2005: Instrumental insemination and Honey Bee Breeding. Short course. - The Ohio State University, Rothenbuhler Honey Bee Laboratory, Columbus, Ohio. 116p.

Cobey, Susan, 2008: NWC stock maintenance protocol. - http://iris.biosci.ohio-state.edu/honeybee/breeding/Evaluation.html

Holm, Eigil, 1984: Artificial insemination of the queen bee. A manual for the use of Swienty's insemination apparatus. - Gedved, Demark.

Honeybee Genome Sequencing Consortium, 2006: Insights into social inects from the genome of the honeybee Apis mellifera. – Nature vol. 443/26.00 oct. p. 931-49.

Jensen, A. Bruun & Vest Pedersen, B., 2006: Honey Bee Conservation: A Case Study from Læsø Island, Danmark. – – In: Beekeeping and Apis Biodiversity in Europe (BABE). ISBN 1-904846-14-9. p. 142-164.

Klug, W.S., Cummings, M.R., Spencer, C.A., 2006: Concepts of genetics, 8.ed. ISBN 0-13-196894-7. 677s + addenda.

Kraus, F.B., 2006: Requirements for local population conservation and breeding. — In: Beekeeping and Apis Biodiversity in Europe (BABE). ISBN 1-904846-14-9. p. 87-107.

Lodesani, M. & Costa, C., 2005: Beekeeping and conserving biodiversity of honeybees. – In: Beekeeping and Apis Biodiversity in Europe (BABE). ISBN 1-904846-14-9. 179p.

Lodesani, M., Costa, C., 2006: Practical aspects of bee breeding for biodiversity aims.

McNeil, M.E.A., 2009: The Bee Chip (om microarrays). - American Bee Journal vol. 149/2 Febr.

McNeil, M.E.A., 2009: Family Reunion. How old World carniolan stock may enrich new world descendants I and II. - American Bee Journal vol. 149/10 and 11.

Norges Birøkterlag: Avlsplan 2005 (med bilag).

Rinderer, Th. E. (ed.) 1986: Bee Genetics and Breeding. - Academic Pres, Orlando. 426 p.

Rua, P., Fuchs, S., Serrano, J., 2006: Biogeography of European Honey Bees. — In: Beekeeping and Apis Biodiversity in Europe (BABE). ISBN 1-904846-14-9. p. 15-52.

Ruttner, Frliedrich (ed.): 1980: Königinnenzucht. - Apimondia, Bukarest. 349 p.

Ruttner, F., 1983: Zuchttechnik und Zuchtauslese bei der Biene. - Ehrenwirth Verlag, München. 141 p.

Schley, P., 1983: Praktische Anleitung zur instrumentelen Besamung von Bienenköniginnen. - Selbstverlag, Polheim. 80 p.

Schlipalius, D. et al., 2008: Honeybee. – Genome Mapping and Genomics in Animals, vol.1, s.1-16. Springer Verlag Berlin Heidelberg.

Solignac, M. et al., 2004: A microsatellite-based linkage map of the honeybee, Apis mellifera. – Genetics 167, p. 253-262.

Solignac, M. & Cornuet, J.M., 2006: Selection theory and effective population size. — In: Beekeeping and Apis Biodiversity in Europe (BABE). ISBN 1-904846-14-9. p.53-86.

Taber, Steve, 1987: Breeding Super Bees. - A.I. Root, Medina, USA. 174 p. NBB (UK) 2009.

Traynor, Kirsten, 2008: Bee Breeding Around the World (about the breeding procedure of Poul Erik Sørensen). - American Bee Journal, vol. 148/2 p. 135-139.

Traynor, Kirsten, 2008: Bee Breeding Around the World. About German Bee Breeding). - American Bee Journal, vol. 148/3 p.237-240.

Vejsnæs, Flemming, 2005: Nordens største dronningavler. (Om Poul Erik Sørensens dronningavl). – TfB 4, p. 112-117.

Index

A
accident 77
acclimatization 77
additive genes 71,74,84
adenine 51
Africa 75
African races 80,81
Africanized bees 54,76
aggressivity 71
A-group 76
aims of breeding 38,40
A-lineage 80
allele 54,55,58,72,89
alphabet 52
American Foulbrood 64,66,84
amino acid 52,71
anaesthetize 16
analysis test 84
anaphase 50
Apis cerana 75
Apis dorsata 75
Apis floreae 75
Apis mellifera 75
ATP 51
Austria 80,87

B
backcross 85
banks of genes 79
base 51
Bienefeld 87
Bo Vest Pedersen 80
box 30
Brazil 76
breeder 49
breeder colony 10
breeder of queens 38
breeding 22,38,79,82
breeding value 87,88
Brother Adam 9,49,79,82,83
Brother Adam's scheme 78
brown bee 58,60,78,83
Buckfast Abbey 82
Buckfast bee 27,79,82

C
cage 8,12
calendar 8
candy 16,17,18,19,20,31
carnica 76,83,87
caucasica 76

cecropia 76
cell 51
cell builder 8,9,10,12,22,23,24
cell cup 10,12
cell finisher 8
centromere 50,59
C-group 76
choice of breeder 44
chromatide 59,67
chromosome 45,46,48,50,51,53,55,
  58,66,72
climate 77
climate change 78
closed population 86
CO_2 17
coding 53
co-dominance 62
codon 52,53
colour 80
colour of the year 15
colza 79
combination breeding 82,83
combination of genes 77
computation 24
computer 44
congregational area 47
container 13
copying error 71
cordovan 26
counting of eaten larvae 70
crossing 39,62
crossing strategy 83
cross-over 46,48,59,66,68,77
cubital index 80
cup container 11
cut wing 16
cutting the wing 15
Cytochrome oxidase I 80
cytoplasm 48,51
Cytosine 51

D
Danish Beekeepers Association
  43,44,84
database 54,87,88
dates for breeding 9
defecate 34
disease resistance 71
diploid 45,46,69
diploid drone 69

DNA 51,53,54,57
DNA-helix 53
dominant 58,60,84
double recessive 65
double-stranded 53
drone 9,20,29,30,31,33,34,37,45,46,
  47,69,72,84
drone congregation area 26,29,46
drone foundation 28,29
drone producers 28
drone production 86
Dronningavlerforeningen 26
Drosophila 54
dry grafting 12

E
ecotype 78,79
egg 45,50,59
egg laying 9
embryo 46
emergency cell 6,7,23,31
English bee 82
enhancer 53
environment 38,44,68,73,74,84
environmental factors 73
enzyme 51,53,54
exon 53

F
F1-generation 83
F2-generation 83
fertilization 45,58
finisher 22,24
fluorescence microscope 53,54,55
foundation 2,19
freezer 66
freezing 87
freezing test 42
fruit fly 54
function of genes 54

G
Gambia 76
genetic burden 73
gene 46,51,52
genetic code 52
genetic diversity 75
genetic drift 56,57,77
genetic examination 80
genetic purifying process 46
genetic research 33

91

genome 45,53,54,89
genotype 65,66,84
gentleness 41,42,71,87
geographic populations 81
Germany 79,87
Glob, A 9
glue 15
grafting 7,8,9,10,11,12,23
grafting needle 10,11
grafting tools 10
guanine 51

H
half sister 47
haploid 45,46,58,69
Hawaii 17,77
heather honey 79
heritability 74
heritage 33,48
hermaphrodite 46
heterosis 73,79,83,87
heterozygote 60,83
Hohen Neuendorf 87
homogenization 86
homogenized sperm 33
homogenizing 33
homologue 58
homozygote 60,73,83,85
Honey Bee Genome 53,75
Honey Bee Sequencing 54
honey harvest 41,90
honey production 42
hygiene 42
hygiene test 43
hygienic 64,65,66,84,
hygienic behavior 41
hygrometer 13

I
Ice Age 75
identical genes 54
inbreeding 33,48,70,73,85,87,88
incubation 12
incubator 8,13,21,22
insemination 33,34,66,87
insemination apparatus 34,35
instrumental insemination 26,33, 34,70,83
insulated 30
intermissa 77
interphase 50,67
intron 53
intuition 44
invert sugar 16

isolated area 56
isolation 77

J
Italian 76
Italian bee 82
judge 44
judgment 38,44

K
Kangaroo Island 79
Kieler nuc 18,19,29,31
killer bees 71
kinship 47,87
kinship, degree of 47

L
Laesoe 60,79
lamarckii 76,77
larva 11
larvae, transportation 10
life expectancy 73
Liguria 76
ligurian bee 79
ligustica 26,60,66,76,79,83
ligustica var. ligurica 82
linkage 66
linked genes 67,68
locus 55,69
Lüneburg 79

M
macedonica 76
maintenance 39,83
maintenance of stocks 85
male 45
management of a nuc 21
Marburger box 31
mark 41,42,44
marker 80
marker genes 92
marking of queens 14,15
marks of the year 43
mated queens 31
maternal 48
maternal genes 47
maternal heritage 33,49,56
maternal set 55
mating 9,60
mating flight 20
mating nuc 18,29
mating on an island 27
mating station 8,18,21,26,30,31,32, 48,55,66,82,87

Mediterranean lineages 80
meiosis 48,58,59,60,67,77
mellifera 26,60,75,76,79,80,81,82
Mendel 62
Mendel's first law 62
Mendel's second law 64,68
Mendelian laws 45
metaphase 50
M-group 76
microarray 54,55,65,89
microclimate 28
microsatellite 55,47,92
mitochondria 48,51,56,57,80
mitosis 50
mixing of sperm 70
morphological characters 80
mortality 33
m-RNA 52,53,56
mucous 29
mutant bees 26
mutation 57,72,89
mutation rates 72
mutations 71
mutations in bees 72

N
nature and nurture 73
newborn 73
nitrogene 42
non-coding 53
nonsense-DNA 53
Nordic bee 76,81
Norway 79,87
Nosema 41,42,74
Nosema test 43
nuc 8,9,14,18,19,20,21,29,30,37
nuc, cleaning 21
nuc, management 21
nucleus 51
nucleus box 8,18
number of matings 47
numbered plates 15
nurse 23
nurse bees 24
nurture 73

O
Oberursel 80
offspring 62
O-group 76
one comb nuc 18,29,30,32
origin 54
origin of races 76
outcrossing 85

ovary 46,48

P
package bees 77
parallellogramme 70
parentel heritage 56
parthenogenetic 68
paternal 48
paternal genes 47
paternal heritage 33,57
paternal set 55
PCR-machine 52
Pedersen, Bo Vest 80
pedigree 27,39,48,49,80,81,86,87,88
penis 29,36,37
pesticide 89
pests 89
P-generation 83
phenotype 65,66,71
pheromone 7,31,47,56,69,77
pin test 42
population 56,62
population genetics 54
Poul Erik Sørensen 22
promoter 52,53
property 78
prophase 50
propolis 76
protein 51,54,71
pure race 79

Q
queen 45
Queen breeders' association 26
queen breeding 26
queen cage 13
queen candy 16
queen cells 12,13
queen container 36
queen excluder 9,10,23,30,31,34,44
queen pheromone 31
queen rearing 7,8
queen right builder 24
queen, abdomen 37
queen, adding 32
queen, age 6
queen, breeding 6
queen, creeping 9
queen, development 9
queen, judgement 43
queen, large scale production 22
queen, number of matings 20
queen, orientation 20
queen, production 6

queen, unmated 18,22
queen, mated 31

R
race 29,51,57,75,76,80,81,82
railway 78
recessive 72,58,84
reproduction 45
resistant 66
restriction enzyme 55
ribosome 51,52
RNA 52,54
robbery 20,32
royal jelly 11
Ruttner 80
Ruttner-Schneider apparatus 36

S
Sahara 77
sahariensis 77
scaffold 29
score 41
scutellata 76
scutellum 14,15
segregation 60,64,66,83,84,85,87
selection 38,83
sequencer 52
sex allele 33,69,70,83,85,86
sex cell 67
sex chromosome 69
sexual reproduction 68
sicula 76
signal molecule 53
sisters 47
somatic cell 45,72
sperm 28,29,33,34,37,45,46,86
sperm cell 50
sperm cells with wings 46
sperm cells, number of 47
spermatheca 69
starter 8,22,23
starting box 25
sting 37
sting, queen 37
stock 82,83,85
stop codon 71
streptomycine 33
sugar water 29
supersedure 7,32
supersisters 47
supporter 22
supporter colonies 23
supporter 24
swarm 6,77

swarm cells 6
swarm trap 77
swarming 32,43
swarming tendency 41,42
Swienty's nuc 19
symbiotic bacteria 56
Sørensen, Poul Erik 22

T
techniques of insemination 33
telophase 50
temperament 41
terminator 53
test cross 65
testicles 45
testing 84
thymine 51
Tracheal Mite 66,82
tranquility 41,42
transport 31
transport cage 14,16,17,20,21,32
t-RNA 52

U
unfertillized eggs 33
uniting colonies 32
unrelated 83

V
Varroa 76,84,88,89
virus 57,92
vitality 73,83

W
wandering 79
wet grafting 11,12
wheat 71
white eyes 72
wing 74
worker 45

Y
yellow 58
yellow bee 60,83
young bees 14

93